AF886189

KVARKADABRA

Sašo Dolenc
THE MAN WHO COUNTED INFINITY
and Other Short Stories
from Science, History and Philosophy

Paperback edition 2018

Cover design by Arjan Pregl
Copy-editing John Stubbs

Published by:
KVARKADABRA - DRUŠTVO ZA TOLMAČENJE ZNANOSTI
Slovenska cesta 15, 1000 Ljubljana, Slovenia
www.kvarkadabra.net
info@kvarkadabra.net

A CIP catalogue record for this book is available from the Slovenian National and University Library.

001(081)
5/6(081)

DOLENC, Sašo, 1973-
The man who counted infinity and other short stories from science, history and philosophy / Sašo Dolenc. - Ljubljana : Kvarkadabra - društvo za tolmačenje znanosti, 2018

ISBN 978-961-94070-9-7

293779200

Sašo Dolenc

The Man Who Counted Infinity

and

Other Short Stories
from Science, History and
Philosophy

CONTENTS

The universe is made of stories, not atoms.

Muriel Rukeyser (*The speed of darkness*)

INTRODUCTION

Albert Einstein once said: "Everything should be made as simple as possible, but not simpler." This is the guiding spirit of the books in this series of "Short stories from science, history and philosophy". The objective here is to explain science in a simple, attractive and fun form that is open to all.

The first axiom of this approach was set out as follows: "We believe in the magic of science. We hope to show you that science is not a secret art, accessible only to a dedicated few. It involves learning about nature and society, and aspects of our existence which affect us all, and which we should all therefore have the chance to understand. We shall interpret science for those who might not speak its language fluently, but want to understand its meaning. We don't teach, we just tell stories about the beginnings of science, the natural phenomena and the underlying principles through which they occur, and the lives of the people who discovered them."

The aim of the writings collected in this series is to present some key scientific events, ideas and personalities in the form of short stories that are easy and fun to read. Scientific and philosophical concepts are explained in a way that anyone may understand. Each story may be read separately, but at the same time they all band together to form a wide-ranging introduction to the history of science and areas of contemporary scientific research, as well as some of the recurring problems science has encountered in history and the philosophical dilemmas it raises today.

NUMBERS

For a moment, nothing happened. Then, after a second or so, nothing continued to happen.

Douglas Adams (*The Hitch Hikers Guide to the Galaxy*)

A mathematical melodrama

On the 30th of May 1832, when the day had barely started to dawn in the southern suburbs of Paris, a young man called Evariste Galois, still drowsy, but with a gun in his hand, stood facing an artillery officer named Pescheux d'Herbinville. A classic duel was about to take place, one where opponents use weapons to defend their honor and reputations. Today, it is not known precisely what the argument was really about, but it seems likely that the main cause of dispute between the two men was the beautiful Stéphanie Félice Poterine du Motel.

From the distance of twenty-five feet the rivals aimed the barrels of their guns at each other and only a moment later the young Evariste fell to the ground, wounded. The bullet hit him in the stomach, puncturing his intestines, and only an immediate operation might have saved his life. However, no doctor was at hand and those watching the duel as well as Evariste's opponent left the scene as soon as he fell. It was only three hours later that the injured Galois was noticed by a passer-by who managed to get him to a nearby hospital. At that particular moment Galois was still fully conscious, but he was aware that his time was fast running out. When his younger brother came to see him, he uttered his last words: "Don't cry, Alfred! I'll need all my courage to die at twenty years of age." The next morning he would not wake up.

Even before this fatal May morning, Galois was well aware that accepting the duel would almost certainly sentence him to death. He was a brilliant mathematician, but hardly as skilled in wielding weapons as his opponent d'Herbinville, the soldier. And so he spent the night before the duel trying to jot down as many of the mathematical ideas that had occurred to him during his short working life, but which he had not had chance to record. We all know that we are usually much more productive when haunted by a deadline than when we have plenty of time at our disposal. However, it is something quite different to be only twenty years old and, in your mind's eye, see the solutions to several mathematical problems which you are certain nobody else knows, while a duel that you are bound to lose is set to take place the very next morning.

That night, Galois filled several sheets of paper with equations and sent them to a good friend, asking him to pass them on to the famous mathematicians Jacobi and Gauss. On these sheets he tried to gather all the ideas that he thought relevant. However, as the surviving notes imply, he found it difficult to focus on mathematics alone. The edges of the pieces of paper contain the words "une femme" (a woman) and a cry of desperation: "Je n'ai pas le temps!" (I don't have time!).

Evariste Galois was undoubtedly a great mathematical genius, but his short life was not without difficulties. Even though he came from a fairly wealthy family - his father was the mayor of a small town near Paris - and received a good education, the trouble was that his teachers, at least when it came to mathematics, could not measure up to his standards. At the early age of fifteen he was already reading - in the original - through every possible paper published by his contemporaries. In order to find a more appropriate study environment, he tried to enroll into the prestigious Ecole Polytechnique, but unfortunately failed to pass the entry exams. When he had already published his first scientific paper later he tried his luck at the Polytechnique again, but was no more successful in convincing the board of admissions than he was the first time.

He also sent his mathematical ideas to be evaluated by the French Academy of Sciences (*L'Académie des sciences*) which was at that time at the centre of scientific work in France and indeed Europe. Galois' work was reviewed by the Baron Cauchy, one of the giants of the French school of mathematics, who praised the achievements of the young mathematician and suggested that he be given a special award from the Academy. Galois developed his ideas further and sent them to the Academy's secretary, but he died shortly after, and the young mathematician's article was nowhere to be found, so he regrettably never received the award even though everybody agreed that he was a most talented young man.

After Galois' sudden death rumors started spreading that the unfortunate incident was not really the result of an honorable duel, but a murder, plotted against the young mathematician because he was thought to be an ardent republican. It is true that he was often present in the first lines of protests and had been arrested on several occasions, but there is no proof to support the theory of premeditated murder. It was actually when serving a prison sentence because of his political beliefs that he made the acquaintance of the fatal Stéphanie. She was the daughter of the doctor of the clinic to which the youngest prisoners were transferred during a cholera epidemic.

Galois' friends and his brother tried to save Evariste's notes after his death, to organize them as much as possible and to pass them on to renowned mathematicians so they could prepare them for publication. However, it took several decades before the world of mathematics came to understand the extraordinary significance of the ideas of this young mathematician.

Anyone who has gone to school in the past couple of decades will remember the unusual mathematical structure called the group. Such knowledge escapes from the head of the average student right after he hands in his last math test, but we should be aware of the fact that modern physics would be quite inexplicable without the use of the group theory. Groups are not only essential to describing the elementary particles which make up the universe; they are a fundamental mathematical tool used by experts from all fields of science. Groups as a basic mathematical structure, indeed one of the foundations of modern mathematics, came into existence on that very fatal night in 1832 when Evariste Galois decided to scribble down equations from his head when he would have done better to practice his shooting.

What is randomness?

You might think it must be easy to define randomness, but nothing could be further from the truth. Not only is it difficult to create random events or sequences of numbers, verifying whether something that we have produced really is random is no easy task either. Many great mathematicians throughout history have examined the problem of randomness, but it was only a short while ago, in the era of computers and information technology, that the questions concerning randomness revealed themselves in all their complexity and appeal.

The paradox of the definition of randomness

It would be easiest to define randomness as a series of events taking place without any meaning or independent of any possible rule. Random is what has neither cause nor meaning. Nevertheless, it is very important not to confuse our subjective unawareness of rules with the objective nonexistence of such laws. We can quite easily come to the conclusion that a certain sequence of numbers is random when we cannot recognize any rule that might govern it, while it is likely that we just cannot make out the pattern. A good example of this is the number pi which represents the ratio of a circle's circumference to its diameter. The definition of pi is very simple, but if we were to see nothing but the long list of decimals with the beginning 3.1415... hidden, it would be extremely difficult to figure out that it is, in fact, a sequence which is far from random.

One could even say that the definition of randomness is, in a way, a paradox. On the one hand, we say that a truly random sequence cannot conceal any rule that would enable us to recreate the sequence, while on the other hand, requiring the absence of all patterns within a sequence leads to a very restricting definition which is almost impossible to apply in practice. For something to be random, it must meet very well defined conditions. Randomness is thus defined by the complete absence of form which is, on the other hand, a very strictly defined form in itself, only with a negative sign.

The shortest possible instruction

In the mid-1960s, the mathematicians Andrej Kolmogorov, Gregory Chaitin and Ray Solomonoff all independently invented a way to effectively define randomness in the era of computers and the digital recording of information. They linked the definition of randomness to the concept of algorithmic complexity. This sounds very complicated, but is actually based on a simple idea.

Andrej Kolmogorov, a renowned Russian mathematician, defines the complexity of a thing as the length of the shortest recipe (algorithm) required to make it. Cakes are usually more complex than bread, because the instructions for making them are normally more extensive. Similarly, orange juice is less complex than beer, for example, as we can reduce the recipe for making the juice to no more than two words: "Squeeze oranges", while the instructions for making beer are much longer.

Following the same pattern, the complexity of a number is defined as the length of the most simple computer program that can write it out. The sequence 01010101010101010101 can be reduced to "10 times 01". We notice immediately that the sequence can be memorized in a shorter and clearer way. When observing a sequence like 01000101000011101001, however, it is not possible to recognize the rule right away, so we have to memorize it as a whole.

If your phone number happened to be 01 1111-111, you would be able to memorize it straight away, just as you would the number 01 2345-678. The rules hidden behind both numbers are simple and we need no more than a single piece of information to recall them. When it comes to more complex numbers, though, we have to commit several pieces of information to our memories.

Sometimes a part of the number contains the date of our birth or a similar sequence that we can easily recall, so memorizing such numbers presents a smaller difficulty than memorizing numbers that possess no apparent pattern. These are, in our eyes at least, completely random.

No recipe for randomness

We can use the concept of algorithmic complexity to define what is random as being a thing for which there exists no shorter recipe than a detailed description of the thing itself. If we limit ourselves to number sequences, then the sequence which is random cannot be reduced to any shorter form or algorithm than the entire actual list of numbers. As there is no rule to sum up the list of numbers, we can only memorize it in its entirety, like a phone number which does not contain a single set of numbers we can recognize.

A binary number, used in the language of computers that can only read ones and zeros, is random when its complexity is equal to the number of its digits. A program that a computer reads in binary code as well cannot be shorter than the number itself. In simpler terms, there is no shorter recipe to write down the number than writing it with all the digits it contains. There can be no other shorter algorithm or program.

All programs for creating random numbers built into our computers are actually only creating the so-called "pseudo-random numbers". The algorithms that create them are shorter than the numbers they produce, so they do not really meet the strict criterion of randomness based on algorithmic complexity. In an article in 1951, John Von Neumann, the famous mathematician, physicist and father of computer science, summarized this problem in a single sentence: "Anyone who considers arithmetical methods of producing random digits is, of course, in a state of sin."

The "Monte Carlo" method

However, modern science could hardly progress without generators of (pseudo-) random numbers. Today, random numbers are most frequently used to enable simulations or to help with more complex calculations. The method of using random numbers to solve mathematical problems is very similar to public opinion polls. If we pose a certain question to a smaller sample of population and if the sample only contains randomly chosen representatives of a population that are not, for example, mostly senior citizens or students, their answers can give us a pretty good idea about the population's opinion concerning a certain subject.

In science, we can deal with the problem of random numbers in a similar way. Suppose we had to calculate the surface of an irregular shape in the form of a heart. Using a procedure that mathematicians named the Monte Carlo method we can evaluate the surface of the heart by circumscribing it with a rectangle, the surface of which is not difficult to calculate. Now, all there is left to do is to evaluate what part of the rectangle is covered by the heart and the problem is solved. But what is the easiest way to assess the ratio of the surface of the entire rectangle to the area covered by the heart? Try randomly placing dots on the rectangle and counting whether you have hit the heart or not. If the dots are really distributed in a random manner, the ratio of the number of dots inside the entire rectangle to the number of dots on the heart will gradually approach the ratio of the rectangle's surface to the heart's surface.

Compressing data

Naturally, verifying whether a long sequence of numbers is random is no easy feat. However, mathematicians have developed several methods for testing different random number generators and checking if the random sequences are good enough to be used for a given task.

In our daily lives, we encounter the process of evaluating the degree of randomness when compressing files on our computers which happens almost every day. As we all know, we can use special software devised to compress data and greatly reduce the size of a file on our computer. These file compressing programs look for repetitions within data and create new dictionaries to write down the information in shorter form. Suppose we repeatedly used the word "problem" in our document. A good file compressor will notice that and substitute all the occurrences of the word with a single sign like *. The sentence: "The problem lies in the problematical approaches to solving problems", will be written in a coded and compressed form: "The * lies in the *atical approaches to solving *s". And the dictionary will add the explanation that the asterisk (*) means "problem".

The following rule applies: the more that the size of the file is reduced by compressing information, the less random data is being compressed. When compressing a written document it is thus easy to evaluate the size of our vocabulary. The more we repeat words in our text, the easier it will be for the file compressor to reduce the size of the file containing our document.

This is why file compressing programs are a relatively good judge of the randomness of a certain sequence. The more the data can be compressed, the less random it is. A truly random sequence cannot be compressed by using file compressing programs because the entire computer algorithm is also the shortest possible algorithm describing this sequence. A file of a truly random sequence cannot be compressed.

When our sense of probability deceives us

The influence of numerous columns published in popular newspaper supplements where so-called "experts" shower us with all kinds of advice is not to be underestimated. These columns do not only form people's opinions and change habits of entire nations; every once in a while they can also provoke large-scale polemics involving wider audiences as well as more professional circles. In the area of health and nutrition such fervent reactions are to be expected, but it is a completely different thing when they are brought about by a mathematical question.

Two goats and a car

Parade magazine, which is distributed as a supplement of more than four hundred American newspapers every Sunday and reaches around seventy million readers, has long featured a column called "Ask Marilyn". It is edited by Marilyn vos Savant who became famous in the 1980s when she was listed in the Guinness Book of World Records for the highest IQ on the planet. In her column, she has been answering various questions and solving problems posed by readers for more than twenty years.

Among all the questions she has ever dealt with, a special place belongs to a seemingly very simple problem which was posed to her on September 9, 1990 by a Mr. Craig F. Whitaker:

> Suppose you're on a game show, and you're given the choice of three doors. Behind one door is a car, behind the others, goats. You pick a door, say #1, and the host, who knows what's behind the doors, opens another door, say #3, which has a goat. He says to you, "Do you want to pick door #2?" Is it to your advantage to switch your choice of doors?[1]

---

[1] "Marilyn vos Savant | The Game Show Problem," http://www.marilynvossavant.com/articles/gameshow.html.

The question became known as "the Monty Hall problem", after the host of the popular American television show *Let's Make a Deal* in which the host Monty Hall challenged his guests to accept or refuse different offers he made. In her column, Marilyn answered her reader that it would definitely be wiser to switch doors because the probability to win the car would increase by two times. This was her answer:

> Yes; you should switch. The first door has a 1/3 chance of winning, but the second door has a 2/3 chance. Here's a good way to visualize what happened. Suppose there are a million doors, and you pick door #1. Then the host, who knows what's behind the doors and will always avoid the one with the prize, opens them all except door #777,777. You'd switch to that door pretty fast, wouldn't you?[2]

Could the most intelligent woman in the world be wrong?

Of course, all of this would have gone unnoticed if it had not provoked a storm of critics aimed at Mrs. vos Savant. *Parade magazine* received more than ten thousand letters sent by outraged readers among which a great many were teachers of mathematics. Nearly one thousand letters were signed with names followed by a PhD and numerous letters were written on sheets of paper bearing the titles of renowned American universities (many of these letters are available on Marilyn's website: www.marilynvossavant.com). All the critics agreed that Marilyn was misleading readers with her answer, because the probability of winning could not be changed by choosing the other door. One professor of mathematics was very blunt: "You blew it, and you blew it big! ... As a professional mathematician I am very concerned with the general public's lack of mathematical skills. Please help by confessing your error and in the future being more careful." Another was so upset that he ended up calling Marilyn herself a goat.

---

[2] Ibid.

The argument quickly spread across the limits of the Sunday magazine and even landed on the cover of The New York Times, and some well-known names from the world of mathematics joined the debate. This is how a reporter described the atmosphere created by vos Savant's explanation: "Her suggestion to switch the doors has been the subject of debate in the corridors of the CIA as well as on Air Force bases by the Persian Gulf. It has been analyzed by mathematicians from the MIT and computer programmers of the Los Alamos laboratories in New Mexico." Apart from the offensive letters which denounced her response, Marilyn did receive some letters of approval. Among others, one came from a professor at the prestigious MIT: "You are indeed correct. My colleagues at work and I had a ball with this problem, and I dare say that most of them, including me at first, thought you were wrong!"

Marilyn was not discouraged by the critics - after all, she was objectively proven to be smarter than all her critics in terms of her IQ – and in one of her following columns she gave teachers across the country the task to play this simple game with their students in class (without real goats and cars, obviously) and send her their results. She later published these results and they were completely consistent with her advice that in the case in question, it was actually wiser to choose the other door.

Who's right?

The debate brought about by the Monty Hall problem falls into the category of what mathematicians call conditional probability. Simply put, this is a branch of mathematics dealing with the question of adapting the probability of an event taking place when new data enters the equation. The essence of the problem which provoked such a widespread and emotionally charged reaction from readers was that most overlooked a crucial piece of information. It is very important to take into consideration the fact that the game-show host already knows which door hides the car.

In the second part of the challenge when the host opens the door which hides the goat, he already knows that there is no car behind this door. If the host did not possess this knowledge and were to open the doors completely at random- as the contestant must- the probability of guessing correctly really would not be increased by changing the decision. In this case, the more than one thousand readers who had concluded that Marilyn was not familiar with the basics of mathematics and sent angry letters to the magazine would have been right. Our mathematical intuition suggests that the probability that the car is behind one or the other door when two of the doors are still closed remains equal. This would of course be true, if there were not for the host who knows more than we do.

This problem is most easily clarified, if we take a look at the action from behind the scenes where we can see which door hides the car and which hides the goats. If the contestant guesses where the car is in his first try, the host will open any of the other two doors in which case it is unwise for the contestant to change his decision. However, this only applies if the contestant chooses the door which hides the car in his first try for which the probability is 1:3 or approximately 33 %. But if in his first try the contestant chooses the door which hides a goat, the host will have to open the only remaining door which hides the other goat. In this case the contestant would be wise to switch the doors because that would certainly win him the car.

If the contestant chooses the goat in his first try it is always to his advantage to switch doors while this does not apply if he chooses the car. The probability that he will choose the goat in his first try is 2:3 while the probability that he will choose the car is only 1:3. If the contestant follows the tactic of switching doors, the probability of winning the car is 2:3; if he refuses to switch doors the probability decreases by half and is only 1:3.

If the contestant chooses to switch doors when the host offers him the chance, he will win the car every time when he chooses the goat in his first try. This choice will win him the car in twice as many cases as it would if he decides not to switch. The probability of guessing correctly increases from 33% to 66%.

If you are having difficulties understanding this right now, do not worry. Many mathematicians needed quite some time to clarify the issue.

Let us conclude with a similar puzzle that you can try to solve yourselves: a family with two children has at least one boy. What is the probability that it also has a girl if we presume that in average an equal number of boys and girls are born? Of course, the correct answer is not 50%.

The man who counted infinity

"There is a concept that is the corrupter and destroyer of all others. I speak not of Evil, whose limited empire is that of ethics; I speak of the infinite." Even though one might expect these words to have been written down by a mathematician or a scientist, this is not so. Their author is an Argentine writer, Jorge Luis Borges, who succeeded in revealing the very essence of a problem that had intrigued many thinkers before him. Infinity is a concept that has appeared time after time in all kinds of different philosophical, mathematical and physical discussions, but has always been clouded by difficulties and contradictions. People simply cannot grasp infinity directly as they do other concepts. We have to imagine infinity in an indirect way. Usually, we describe infinity as an endless or limitless sequence, but we quickly come upon problems, as with infinity we enter a dimension where our intuitions may not be trusted completely.

There is no single infinity

Throughout history, a great number of thinkers and scientists have pondered the concept of infinity and reached many interesting and important conclusions, but the greatest development in understanding this difficult idea occurred in the second half of the 19th century, when the mathematician Georg Cantor (1845-1918) approached the subject from a completely new angle.

With a couple of simple definitions Cantor succeeded in setting new boundaries to this complex area of thought, which allowed him to examine the concept of infinity thoroughly and systematically. To the great amazement of the scientific and philosophical community he demonstrated that there is no single infinity, but, in fact, an infinite number of different infinities. He came to this conclusion after careful and accurate consideration and was also able to prove formally all the stages leading to his conclusion His ideas subsequently became incorporated into the very core of modern mathematics.

One of the first challenges he had to overcome was defining what infinity actually was. Of course, the simplest way to define infinity was to describe it as having no limits: infinity is what is larger than anything finite. This is where Cantor made an important step forward: for he defined something as infinite, if some of its parts are as big as the whole. This might sound contradictory as we are all intuitively used to finite dimensions whose parts are always smaller than the whole, but as we have already mentioned, when it comes to questions of infinity, intuition does not offer the best answers.

How to compare infinities?

Not only did Cantor invent new ways of dealing with infinity, he also came up with the well-known set theory that children have been learning in elementary school for decades. When thinking about his new theory, he soon had to face the question of how to compare the size of two sets. He proceeded from the completely intuitive supposition that two sets were of the same size if each element from the first set could be matched-up with exactly one element from the second set. Two groups of children are of the same size if a child from one group can hold hands and pair up with a child from the other group.

Cantor generalized the method of comparison by "holding hands" to infinite sets. Two infinite sets are of equal size if we can match-up each element from one set with an element from the other set. According to this definition, the set of even numbers is just as big as the set of all natural numbers since we can pair up each natural number with an even number simply by multiplying it by two. One holds hands with two, two holds hands with four, three holds hands with six and so on all the way to infinity. As we have matched up each element from the first set with an element from the second set, both of the sets, according to our definition, are of the same size, even though we could also say that the entire list of natural numbers is twice as large as the list of all even numbers, because natural numbers are made up of both even and odd numbers.

It was with the help of this strange notion that two sets were of the same size even though one is twice as large as the other that Cantor defined the infinite set. An infinite set contains a subset of the same size, when size is defined according to the principle of holding hands. Natural, even and odd numbers thus represent infinite sets.

How to count points on a line?

The question that immediately arises is whether all infinite sets are of the same size. For example, is a set of all points on a line as big as the set of natural numbers? In other words: is it possible to count all the dots on a line? One of Cantor's important achievements is the proof that it is impossible to count all the dots on a line. Let us have a look at his proof, also called the diagonal argument.

Suppose we are counting points on a line which is one unit long. Each point is matched up with a number between zero and one. As points fill out all the line because there can be no space between them, otherwise we would not have a line, but a sequence of dots, all of the points on a line can only be matched up with numbers if we use numbers with an infinite decimal representation, called real numbers. Every point on a line which is one unit long can be matched up with a real number between zero and one.

Now imagine we wrote down all of the real numbers one under the other in an endlessly long list on which every line contains one real number and the lines are marked with natural numbers. Cantor proved that we can always find at least one real number missing from every such list of numbers. How come? We simply take away the first decimal from the first real number on the list and change it, then take the second decimal from the second number and change it … If we continued to do so until we came to the end of the list, we would create a new real number different from any number already written on the list.

Coming back to our analogy with the two groups of children that we have compared in size by matching the children up in pairs, we have now proven that pairing up natural numbers with points on a line or real numbers never works. We can always find one real number or point on the line left that has no pair.

Infinities come in infinite numbers

The diagonal argument can also be used to prove that the set of all subsets of a given set is always larger than the set itself. It is the easiest to do so with natural numbers. We write down the subsets of natural numbers as lists of on and off numbers in the set of natural numbers. Odd numbers, for example, are written as {1,0,1,0,1,0 ...}, and prime numbers as {1,1,1,0,1,0,1 ...}. Next we arrange these sets into a long list and apply the same argument as before to show that we can always construct another subset of natural numbers that has not been on the list before, simply by taking one element from each subset and changing its value. The set of all subsets of an infinite set is larger than the set itself. Infinite sets come in infinite numbers.

However, a problem already occurs with the smallest infinite sets. We know that natural numbers are the smallest infinite set. It still remains unclear though which is the next largest infinite set. Is it the set of real numbers or points on a line? Or is there another infinity in between, bigger than natural numbers and smaller than real numbers? Cantor presumed that such a set does not exist, but he was unable to prove it. Many years later, mathematicians solved this question not by finding the answer, but by showing that the question had no answer at all.

Many books about Cantor's efforts to solve problems concerning infinity also mention his illness which caused him to spend the last years of his life in a mental hospital. Today, he would have been diagnosed with bipolar disorder or manic depression, but at the time most people suffering from this illness were simply labeled insane. Many writers have implied it was actually his work on infinity that drove Cantor over the edge of sanity. This might sound like a reasonable theory, but it is unlikely that his illness and his research were directly related.

The hermit of the Pyrenees

In August of 1991, Alexander Grothendieck, generally regarded as one of the most important mathematicians of the 20th century and whose influence is often compared to that of Albert Einstein, suddenly left his home in the south of France and headed for the Pyrenees. Since then, he has been living as a hermit high in the mountains somewhere between France and Spain, completely cut off from civilization. In the mid-nineties, a few mathematicians managed to reach him at his home in the wilderness, but for the last couple of years he has remained unseen. His mail is still piling up at the University of Montpellier, but he explicitly prohibited even the handful of his friends who, at the beginning, knew where in the mountains he lived, to bring it to him. Today, even his closest relatives are not completely certain if he is still alive.

Even before his departure into the deep wilderness, Grothendieck lived a very secluded, ascetic life in an old house with no electricity in a village near Montpellier in France. After a successful mathematical career in the 1950s and '60s, when he was also one of the principal members of the infamous Bourbaki group, he became increasingly interested in ecological and anti-war political movements in the 1970s. He became so involved with the struggle for social justice that he traveled to Vietnam in protest, participated in numerous demonstrations and even went took his ideals so far as to refuse a national research scholarship, in order to avoid even tacitly supporting a system of national politics he utterly opposed.

Merely to begin understanding Grothendieck's complete devotion first to mathematics and then politics and ecology, one must look back to his childhood. His father Sasha was a convinced anarchist and participated in several rebellions in imperial Russia at the beginning of the 20th century. In 1921, he moved from Russia to Berlin where he moved in radical circles and met his future partner, Hanka, who came from a wealthy bourgeois family, but associated with members of avant-garde movements. They had little Alexander on March 28, 1928. At the time, the young family also supported Hanka's daughter Maidi from her first marriage.

In 1933, when Nazis came to power, Alexander's father Sasha fled Berlin for Paris and was soon followed by Alexander's mother, but she did not take her son and daughter with her. She placed Alexander into foster care with a family that lived near Hamburg, and left her daughter in an institution for handicapped children, even though she was a perfectly healthy child. Alexander lived with his foster family from his fifth to his eleventh year. He rarely received letters from his mother and never even heard from his father nor from his other relatives who lived in nearby Hamburg. Naturally, this period of separation from his parents left a deep mark on young Alexander.

In 1939, the political pressure became too great and the foster family could no longer take care of all the children. The problem with Alexander was that he looked distinctly Jewish, an appearance with ominous implications not only for him but also the rest of the family. So his relatives found Hanka with the help of the French consulate, sat little Alexander on a train and sent him from Hamburg to Paris. Both of his parents had spent some of the years away from their son in Spain, where they fought against Franco. On his return to France his father was arrested as a "dangerous foreigner" by the French authorities of the time and sent to an internment camp. He died a few years later in Auschwitz.

Hanka in Alexander spent the war in different internment camps, but as soon as the war was over he enrolled at university to study mathematics. He was not impressed with his teachers, so he mostly studied on his own. Before his twentieth birthday, much like young Einstein, he independently made several important mathematical discoveries, unaware that they had already been made and published by other mathematicians.

When he moved to Paris he started to spend time with the most prominent French mathematicians of the time and joined the Bourbaki circle of which he quickly became a driving force. He was becoming more and more famous for his highly abstract approach to solving mathematical problems. His friends later claimed that he was unable to think about concrete things, because his mind only functioned on a universal level.

After a long and productive collaboration with the Bourbaki, he left the group in protest, because most of the members refused to accept his suggestion to replace the old set theory for formalizing mathematics with the new more general category theory which he had helped create. The set theory was constrained by a number of paradoxes and so it became too narrowly oriented to be appropriate for describing the entire diverse scope of modern mathematics. The mathematician Pierre Cartier, one of the more important members of the Bourbaki group summed up the essence of the problem: "Set theory is too restrictive: an element is either a member of a set or not, there is nothing in between."

The decision of the Bourbaki to refuse Grothendieck's suggestion to move away from the set theory to the category theory was, as it soon turned out, a big mistake. Category theory became a very important area in mathematics in the years to follow, and Grothendieck received many awards for his achievements, among others the Fields Medal, also known as the "Nobel Prize of Mathematics".

Is marriage a mathematical operation?

One autumn Sunday in 1934 the director of the prestigious Parisian school *École Normale Supérieure* (ENS) called up a young philosopher called Claude Lévi-Strauss and asked him if he might be interested in applying for the position of Professor of Sociology at the University of Sao Paolo in Brazil. Lévi-Strauss, who had not yet reached his thirties, accepted the offer to leave for this faraway place mainly because he wished to distance himself mentally from the European intellectual scene. He thought it to be too involved in dealing with abstract problems and was more interested in immersing himself in anthropological field work.

A World on the Wane

Lévi-Strauss later remembered that the director of the ENS was, at the time he offered him the job in Sao Paolo, certain that natives of the country's interior were already to be found in the suburbs of the Brazilian metropolis, and that the philosopher, interested in the relatively new discipline of gathering information about peoples of different cultures and their customs in the field, could study their culture during weekends.

Naturally, there were no natives near the university where Lévi-Strauss spent the following couple of years lecturing. On several occasions, however, he did set off for more remote places in the rainforest where he could learn about the customs and way of life of the Indian tribes. Even though he spent a lot of time in the field and is today considered one of the principal names in twentieth-century anthropology and philosophy, he was neither a classical field anthropologist nor a classical philosopher who almost never leaves his desk.

In fact, he was interested in solid data describing the customs of specific cultures and wanted to use the information he gathered to reveal the universal structure, characteristic not only of a certain tribe, but of all human societies. He later described his expeditions to the Amazon region in detail in his book entitled *A World on the Wane*, which combines his autobiographical notes with an analysis of the lives of Indian tribes.

Are you the guy who makes jeans?

On his return from Brazil he quickly realized that the situation in Europe was even worse than when he had left for South America. The air was thick with the anticipation and anxiety as the Second World War drew near, and because he was of Jewish origins, he only thought it wise to leave France once again. He decided to accept an invitation from the renowned private institution, *The New School for Social Research* in New York, which offered him tenure.

Immediately after his arrival to the US, his acquaintances kindly suggested that he change his name as soon as possible. If he refused to do so, anyone who met him would think that he sold jeans. Even though it all seemed to be a joke at first, he quickly came to realize that the similarity of his name to that of the famous jeans manufacturer was more of a nuisance than an entertaining coincidence. In order to avoid unwanted misunderstandings he decided to sign his name Claude L. Strauss during his stay in the US. However, this change was not effective enough to save him from receiving orders for the legendary blue jeans to his address every once in a while.

The French in Manhattan

In New York, Lévi-Strauss made the acquaintance of the Russian linguist Roman Jakobson who also lectured at The New School. They quickly discovered they had very similar approaches to science. Lévi-Strauss tried to introduce the method that Jakobson had been developing in the field of linguistics to anthropology. In order to become familiar with each other's past research they also frequented each other's lectures.

Many other French intellectuals spent the wartime years in New York. During the week, Lévi-Strauss spent his time with Jakobson, immersed in problems concerning anthropology and linguistics, while at weekends he joined the painter Max Ernst and the writer André Breton to look for Native American art in the local markets.

The Elementary Structures of Kinship

Jakobson was very impressed with Lévi-Strauss's doctoral thesis so he encouraged him to publish it in book form as well. In his thesis, Lévi-Strauss examined the structures of family relations in different cultures around the world. Each culture has its own specific rules determining who can marry whom, and determines which newly created family bonds are strictly forbidden.

The Elementary Structures of Kinship, as the book was called when it was finally released after the Second World War, is still considered to be one of the most important works in the field of anthropology. It also had a great influence on other fields of science.

When Lévi-Strauss was preparing his manuscript for publishing in New York, however, he realized that he was still missing the essential part. He had plenty of data which he had gathered in the field, but he wanted to arrange it into a coherent whole, and in this he hit an obstacle. While using the same approaches that Jakobson had been applying to linguistics, he failed to find the inner logic within his field data on the acceptable and prohibited ways of creating new family relations.

As he knew that finding a structure within data was also a mathematical problem, he turned to his mathematician colleagues, but they were not of much help. One of the representatives of the older generation even advised him to stop looking for principles because he would not find any: "Mathematics only knows four operations and marriage isn't one of them."

A mathematician comes to the rescue

After a while, he finally found some good fortune. In New York, he met a young French mathematician, André Weil, who also gave lectures at one of the American universities. Weil was a mathematician of a new breed who did not have much in common with his elderly colleagues. He was one of the founders of a group of young French mathematicians who published their scientific discussions and university textbooks under the collective pseudonym Nicolas Bourbaki.

Their basic principle was to set mathematics on a new footing, mostly with the aid of set theory. In fact, The Bourbaki group was responsible for the famous new mathematics reform, moving the focus from arithmetic to sets and operations involving sets.

In his research, Lévi-Strauss gathered a large quantity of data on family relations in different cultures of the world. He found out, for example, that similarities exist even between family structures of groups living as far apart as the Indian tribes of Brazil and the Aborigines of Australia. However, he could not succeed in revealing the general structure or system of all the acceptable newly formed family bonds.

"When in doubt, look for groups!"

After Lévi-Strauss explained his problem to him, André Weil immediately suspected that a structure that could organize the mass of anthropological data into a coherent whole could be a mathematical group. This completely abstract algebraic structure, which was not as well known in the mathematical community then as it is today, turned out to be one that can also be found governing the most unlikely situations, such as marriages between Australian natives.

Weil was well known for his motto: "When in doubt, look for the group!" And this abstract algebraic approach was indeed useful when applied to anthropology. The essential principle that

Weil applied to Lévi-Strauss's data was that he ignored the very elements among which he was supposed to seek out regularities. He did not focus on the types of marriages, but on the relations between weddings. The structure was not hidden in the marriages as such, but in the differences between types of marriages. This approach was in complete accordance with the ideas of the Bourbaki group that relations and structures are the central elements of mathematics.

Weil found out that the structure of relations between the marriages of the members of different generations and tribes was defined by what he called a permutation group. He described his findings in a discussion which was published as an appendix to Lévi-Strauss's influential work. More than a concrete analysis of family relations, this book is important because of its influence, as it introduced the notion of structure as a system of differences, causing a veritable intellectual revolution.

The interdisciplinary cooperation of Lévi-Strauss, Jakobson and Weil in New York in 1943 became a thing of legend. Three leading experts on anthropology, linguistics and mathematics, each brilliant in their own field, so different at first glance that it is hard to imagine how they could talk about anything other than the weather, came together to achieve an amazing scientific breakthrough. According to science historians, this was when the structuralist movement, which influenced all social sciences, mathematics and philosophy in the second half of the 20th century, came to life.

A mathematical intrigue at the Swedish court

Monetary prizes for solving complex scientific and technical problems are not awarded as often as they used to be. Today, tenders where a committee of experts offers prizes for the best creations sent to its address by a certain date are a common way of finding solutions in the fields of architecture and other artistic/technical disciplines, but not so much in science. Every once in a while an organization or an individual still announces that a prize will be given to anyone who can solve some seemingly insoluble problem, but these awards are not really significant in terms of maintaining or increasing the financing of scientific work. In the past, however, this used to be quite different.

A little more than a hundred years ago, the Swedish mathematician Gösta Mittag-Leffler persuaded King Oscar II of Sweden to organize a mathematics competition in honor of his sixtieth birthday. As the king was a student of mathematics himself, he took to the idea quickly, especially because Mittag-Leffler had succeeded in convincing two highly distinguished European mathematicians to take part in the jury that would set the tasks and evaluate the solutions. The invitation was accepted by Karl Weierstrass of Berlin and Charles Hermite of Paris.

Is the solar system stable?

Four tasks were given, all relating to problems that were also principal subjects of mathematical research at the time. Even though the king's birthday was not until 1889, preparations for the award already started in 1884, so that scholars from across Europe would have enough time to examine the problems thoroughly and come up with their solutions. In the middle of 1885 *Nature* magazine printed an announcement inviting scientists to participate in a mathematical competition in honor of King Oscar II's sixtieth birthday. The deadline for submitting the solutions to the Swedish court and the president of the jury Gösta Mittag-Leffler was June 1, 1888. Naturally, a precondition of entry was that all submissions be made anonymously, to guarantee unbiased evaluation of the work. That is why all candidates were required to enclose with their solutions a sealed envelope containing their names.

The first of the four questions which were given in formal mathematical language pertained to the problem of the movement of three bodies. In other words, the jury was interested in the question of the stability of our solar system. Today, more than a hundred years after the event took place, what happened a couple of months after the announcement of the winner is more interesting than the initial question and its answers. When the winning solution was already being prepared for publication the editor of the magazine noticed that the jury had rewarded a work which was not completely flawless. One of the mistakes actually turned out to be of key importance.

The winner was no surprise

The winner of the competition was none other than the eminent French mathematician Henri Poincaré who, despite his young age at the time, already enjoyed a very high reputation. In the period between the announcement of the competition and the deadline for submissions he was even elected a member of the French Academy of Sciences, a great honor for a man of only thirty-two.

The competition was not about winning the money. The awarded sum came nowhere close to that of today's Nobel Prize. The winner received 2500 Swedish kronor which was approximately one third of a professor's yearly salary. The award improved the young mathematician's scientific career significantly and enabled him to secure a good position at a renowned university, but it certainly did not make him rich. As we shall see shortly, the prizewinning Poincaré actually lost more than he gained from the award.

With his 158-page-long solution to the first problem Poincaré enclosed a short accompanying letter which he signed, even though all of the submitted essays were supposed to be anonymous. During the evaluation process the jury thus already knew who had written the solution which they later voted unanimously voted as the best. The winner was announced during the celebration of the king's birthday on January 21, 1889. A part of the prize was also the publication of the solution in the prominent mathematical journal *Acta Mathematica.*

It was at this moment that the problem occurred. In July 1889 when the editor of the journal was preparing Poincaré's article for publication he noticed a few minor mistakes in the manuscript. He notified Mittag-Leffler who, in turn, wrote a letter to Poincaré on July 16 to inform him that all of the mistakes but one could be fixed immediately. The one mistake was neither a typing error nor a minor mathematical error. It turned out to be of a much greater importance. As it soon turned out, Poincaré overlooked something essential in his proof which was only discovered after the prize had already been awarded and the article was almost in print.

How to avoid a scandal?

What to do? If it were to become known that a flawed solution had been awarded the prize, the reputation of both the king and all the highly esteemed mathematicians who took part in the evaluation process would be badly tarnished. The disgrace would be especially devastating to Poincaré, who had just begun to shine as the new star on the European scientific scene.

Poincaré went back to work in order to correct the mistake, but the more he examined the problem the more it seemed it was not merely a minor error that could be fixed easily. At first, he sent extensive comments to the text which were meant to clarify the issue, but it became increasingly obvious that these explanations alone would not be sufficient to deal with the difficulty.

So what kind of an error was it? In his original solution Poincaré used a completely new method that made his work much easier and brought about a minor revolution in the solving of similar mathematical problems. Instead of calculating the entire orbits of individual bodies or planets, Poincaré chose to focus only on specific moments when a planet or asteroid intersects with a chosen plane. In other words, this method would be best compared to taking an image of the solar system at the exact moment when an observed body circles the sun. In the case of Earth, this would happen once a year.

Poincaré only wanted to determine if a planet or an asteroid, after circling the sun, returns to the same spot, or to find out how the yearly positions change with time. In slightly more technical language: if a chosen plane is always intersected at the same point, the orbit of the motion is obviously stable, but if the plane is crossed at a slightly different point each time, this change has to be described and examined to determine whether these deviations occur within a stable system.

The chaotic mechanics of the stars

Poincaré's discovery, which he had also argued for in his original essay, was that the solar system was stable, at least in a simple model of the sun with one bigger planet and one small asteroid orbiting in the same plane. His first finding was that, in this case, the orbits of motion were stable. When he examined the problem again, it became evident that he had forgotten to take into account an entire range of solutions which were not stable, but would lead to chaos.

It turned out that Poincaré had forgotten to take into consideration a geometrical configuration which leads to completely different solutions than those he had described in his original essay. Today, we would describe these new solutions, which he overlooked at first, as chaotic. Even though they are precisely determined by clear equations and the paths of the bodies in question are such that it is possible to predict their future motion, this kind of prediction depends on the knowledge of the initial conditions. If there is only a slight difference in the position of the bodies at the beginning, they are bound to move in a completely different way. On the basis of this finding Poincaré could only conclude that not all forms of movement in a simplified system of three bodies were stable, which also meant that the solar system or any other similar system is not necessarily or absolutely stable.

In the couple of months Poincaré spent striving to correct the mistake he had made in his original article, he set the foundations for a new branch of mathematics which later developed into chaos theory. The hundred and fifty-eight pages of the original essay expanded to two hundred and seventy. The problem, however, was that the original article had already been printed in Sweden. Fortunately for all those involved it was not yet distributed to the subscribers. To avoid a scandal Poincaré paid 3585 kronor for a corrected reprint of the issue, much more than the sum of the prize money he had received. The organizers also did their best to ensure that the first prints of the journal were destroyed, but a couple of the original issues survived. In 1985, during a research visit to Sweden, the American mathematician Richard McGehee was looking through some archives and came upon a couple of copies of the thirteenth issue of *Acta Mathematica* in an unlabeled box containing Mittag-Leffler's letters. On looking through the contents McGehee discovered that Poincaré's article was different to the one he was familiar with from other copies of the same journal.

Even though Poincaré won the prize for the first, flawed version of the article, the edited article, with which he made a giant leap towards the science of chaos, is much more important from today's point of view. It might seem self-contradictory, but the award was definitely given to the right person, although for the wrong solution and the wrong essay.

Worshippers of mathematical infinity

At the beginning of June 1913 a Russian naval fleet attacked Mount Athos in Greece, a place of worship that has long been the spiritual centre of Orthodox Christianity. Several hundreds of monks lived in monasteries all across the Athos peninsula. The Russian sailors who came ashore were ordered to gather the monks in the monasteries' courtyards. As many of the monks barricaded themselves inside their cells, the soldiers were obliged to use force. The monks of Athos had been autonomous even during the Turkish reign, so the Russian military offensive was more than unusual.

The reason for this encounter was the spreading of a sect worshipping the name of God. Its members practiced a special meditative form of prayer during which they pronounced the name of God in order to become one with God. The founder of this doctrine was the monk Ilarion who was said to have achieved a state of religious ecstasy and closeness to God through this form of meditation.

The sailors were accompanied by the Archbishop Nikon who came to explain to the gathered monks why the Church believed that worshiping God's name was a form of heresy. In a formal examination, a special commission of the Russian Orthodox Church determined that this was a heretic sect because its members equated the essence of God with his name, which meant that they had confused the name with what it stood for. The monks then had to declare their allegiance. According to official records, 661 declared themselves against the so-called heresy, 517 did not concede and were labeled as heretics while 360 refused to make a choice. Roughly a thousand heretic and undecided monks were then taken back to Russia on the ships. There, they were interrogated by the police and forty of the most passionate worshippers were sent to prisons, while the remaining monks were forbidden to return to Mount Athos. The sect continued with its activities in secret, which allowed it to survive even after the October Revolution and the creation of the communist regime, since it had already been underground during the time of the Monarchy.

Mathematicians and mystics

It is interesting that several highly prominent Russian intellectuals belonged to the sect. Among others were some of the finest Russian mathematicians of the early twentieth century, who had established the renowned Moscow school of mathematics, an institution which was at the forefront of research and would produce some of the most influential mathematicians of the coming century.

The three key figures who had joined the sect and were also leading mathematical researchers were Dmitri Egorov, Nikolai Luzin and Pavel Florensky. Egorov and Luzin were professors of mathematics at Moscow University while Florensky was mostly a philosopher and theologian, even though he had studied mathematics and was involved in scientific research.

Egorov and Luzin kept close ties with other European mathematicians who they visited on a regular basis, so they were always up-to-date with the principal problems of mathematics, which was revolutionized at that time by new discoveries about different concepts of infinity. As we have seen, the existence of different infinities had been proved by the German mathematician Georg Cantor with the aid of another new theory, the set theory.

In their book *Naming Infinity: A True Story of Religious Mysticism and Mathematical Creativity* (Belknap Press of Harvard University Press, 2009), historians of mathematics Loren Graham and Jean-Michel Kantor try to illustrate how much the mathematicians' affiliation to the meditative sect influenced their scientific work, especially when they came to deal with the problems of infinity raised by modern set theory.

Russian mathematicians of the late nineteenth and early twentieth centuries frequently linked their mathematical research with religious and philosophical questions. That is why it is not surprising that the theological issue of naming God was associated with problems stemming from set theory. Like God, infinite sets only become real when we name them. The creative power of naming was exercised by equating the construction or definition of a new infinite set with its creation. Mathematics generates a new entity or set by naming it or showing how it differs from others. According to some interpretations, these mathematicians believed the ultimate set to be God himself.

The Moscow school of mathematics

Directly before the start of the Second World War, Luzin and Egorov initiated a new postgraduate seminar at Moscow University, which later became the famous Moscow school of mathematics. The seminar quickly became known as Lusitania, and though it remains unclear where the name came from, it was most probably derived from Luzin.

The narrow circle of seminar members was a kind of secret society. Another peculiarity was that some of the members were still very young. Andrey Kolmogorov who later became a very famous mathematician joined the group at seventeen years of age, while some of the other members were no older than fifteen. The seminar was also frequented by female students, the most notable of which was Nina Bari, who joined the mathematicians when she was seventeen and went on to become the first woman to obtain a doctoral degree from Moscow University.

Lusitania was also known for its cold lecture rooms, which were more often than not left unheated. The dean of Moscow University would only allow classes to be cancelled if the temperatures inside dropped below -5° Celsius. Nevertheless, the seminar students were so passionate about mathematics that they insisted on staying in, even when the temperature was lower. They would come to class wearing heavy fur coats and attentively watched each other's faces. If someone started to show signs of frostbite, white spots on the skin, others would rub them to warm them up. To get their bodies moving and keep warm, they carried snow to the hallway in front of their classroom and drenched it with water, creating a veritable ice rink on the second floor of the building.

However, the greatest danger this group faced was not the cold or the hunger, but political purges that swept across the halls of the university and various institutes. The mathematicians were lucky because they needed no more than a pencil and a piece of paper to do their work, and were, in any case, dealing with questions that no one else could understand. The mathematicians wrapped themselves in a cloak of silence and continued to develop theories in their ivory tower, managing for quite some time to avoid the purges. When communists seized power, Luzin and Egorov no longer mentioned religion to their students, but continued to maintain in their approach to mathematics the philosophy they had developed through religious analogies.

Mathematics in the times of Stalin

However, there were some mathematicians among militant Stalinists who had an idea about what Egorov, Luzin and their young students were doing. Ernest Kolman was a very well educated scientist who had studied abroad and written several influential books on the history of science and mathematics, but turned out to be a real monster during revolutionary times. He used various Stalinist methods to put pressure on both scientists. Egorov was soon imprisoned on charges of associating mathematics with religion and taking part in an antirevolutionary organization. Because he was not allowed to perform his meditative prayers in prison, as other members of the heretic sect did, he went on a hunger strike and died of malnutrition.

Luzin was more careful in concealing his religion and presented a much tougher challenge to Kolman. In 1936, however, he was exposed due to a banal mistake. In order to please the new authorities he visited different schools and lectured on mathematics. On one of these visits, Kolman sent a reporter to ask Luzin to write a newspaper article describing the event. In his article, Luzin praised the students, saying they were very good mathematicians, and this almost cost him his life. Kolman summoned him for questioning because he knew that, in truth, the students' knowledge was very poor. Luzin was accused of using false praise to weaken the Soviet education system.

The academy Luzin belonged to had to form a special commission to examine his work. Even though it was obvious that Luzin would be sent to prison and would most likely end up being shot, the situation took an unexpected turn. Stalin himself intervened and had him liberated. It was only revealed half a century later that Luzin's release was due to the efforts of the physics Nobel Prize winner Peter Kapits. Kapits sent a confidential letter to Molotov who then gave it to Stalin. In it he wrote that he was not familiar with the charges against Luzin, but that he thought it would be a shame if Russia were to give up such a talent. He also wrote that Newton was a religious fanatic, but it would still be a great shame if he had been incarcerated. Altogether, Kapits wrote more than a hundred such letters to Stalin, Molotov and other notables. Surprisingly, many of his requests were granted.

The man who believed machines could think

In 1952, the mathematician Alan Turing called the police to report that his house in Manchester had been burgled. The thieves did not take much, so it first seemed that the official statement of the burglary would be no more than a formality. It was not very likely that the burglars would be discovered and the stolen items and money retrieved. It soon turned out that the mathematician would have been much better off if he had not contacted the police. After a brief questioning on the possible background of the burglary, Turing was suddenly transformed from the victim into the criminal.

A moral crime

Turing confessed to the officers that he had been seeing a nineteen-year-old youth called Arnold Murray who he had met shortly before the incident. Arnold had spent a couple of nights at his place and also borrowed some money, so it was very possible that he had something to do with the burglary. Now, the police were no longer interested in the details of the theft, but in another offence which Turing was careless enough to confess.

At the time, homosexuality was still considered to be a crime in England and the police quickly grasped the true nature of the relationship between the two men. Turing was summoned to appear in court on the same charges ("gross indecency") for which Oscar Wilde had been sentenced to two years in prison half a century earlier. During the trial Turing never denied his "crime", and was - according to the law from 1885 which was originally created to protect women - sentenced to choose between imprisonment or compulsory treatment of homosexuality. He opted for a course of treatment which was more like torture and did nothing to change his sexual orientation.

In order to avoid prison he had to, in accordance with the sentence of the court, undergo a one-year therapy with the female sex hormone estrogen which was supposed to eradicate his lust for men. For a year, he received regular shots of estrogen which caused him to gain weight and visibly develop breasts. In reality, the hormone therapy was a sort of chemical castration as it did not "heal" his homosexuality, the worst part of it being that it also affected his ability to think and concentrate. Because of this completely absurd law, the British legal apparatus systematically destroyed one of the greatest minds of England. At the time Alan Turing was just over 40 years' old.

100 km bicycle ride to school

In the beginning of the twentieth century Turing's father served as a British colonial officer in India. Because his mother believed the Indian environment to be inappropriate for the education of her two sons, Alan and his older brother John spent their youth in different English institutions for children and, later, in boarding schools. In 1926 when Alan was 14 years old, he was accepted by the renowned Sherborne School, but almost came in late for his first day of class. England was just then in the middle of a general strike during which public transportation was cancelled, so young Alan had to set off on a hundred-kilometer journey from his home to the school on a bicycle. His unusual cycling exploit was even mentioned in the newspapers.

He continued to lead an athletic life, especially as a long distance runner. He often used to run to science conferences, sometimes even beating his colleagues who preferred less demanding means of transportation. He would actually have been selected for the 1948 British Olympic team had it not been for an unfortunate injury.

Alan discovered his homosexuality early, during his teen years, when he was strongly attached to one of his classmates who suddenly died after complications in tuberculosis treatment. The loss of this good friend in his youth also dealt a blow to his religious faith. He became an atheist and firmly convinced that everything had to have a factual explanation, even the processes which lead to thoughts in the brain.

After failing twice to win a scholarship to Trinity College at the University of Cambridge, he decided to enroll into King's College where, among others, John Maynard Keynes and E. M. Forster lectured. After writing an excellent doctoral thesis, he was invited to remain at the university as a teacher and a researcher. If it had not been for the Second World War, he probably would have continued his academic career for a long time, but his country needed him for another, highly important national security project.

How to decipher the Enigma?

On obtaining his doctorate Turing left for the United States and spent a couple of years at Princeton University, but returned at the outbreak of war to join the secret decryption centre at Bletchley Park, situated between Cambridge and Oxford. This was where the British gathered mathematicians, chess masters, Egyptologists and all others who might be in any way capable of deciphering intercepted German military messages.

It was particularly difficult to decipher the enemy's coded messages because the German Army used a special mechanical device called the Enigma which looked like a kind of typewriter. With the aid of this device they could encode their messages so efficiently that they were certain that, without the decryption key, which they changed every day, nobody would be able to read them.

Soon after his arrival to Bletchley Park, Turing and his colleagues developed an electro-mechanic machine which enabled them to break the German code each day and read the messages of the hostile army. It was especially important that they also succeeded in breaking the additionally reinforced code system which had been used by German submarines in the North Atlantic. Turing's ideas, in which he combined all of his previous research in the foundations of logic and mathematics with the notion of a machine that could do great amounts of work quickly and efficiently, were of key importance in the successful deciphering of the Enigma.

Like many great scholars, Turing was a peculiar character himself. He rode a bicycle to his post in Bletchley Park wearing a gas mask which he believed would protect him from pollen. In the tearoom he chained his cup to the radiator, so that nobody would take it.

Until the British removed the label *classified* from the documents containing information about their great wartime project of decoding German military messages, Alan Turing was best known to the public for his achievements in the fields of mathematical logic and the theoretical basics of artificial intelligence. It was only after this revelation that it became clear how important his influence had been on the course of the the war.

The poisoned apple

After the war, Turing returned to Cambridge hoping to find some peace in his academic work. He joined a group that was trying to build a working prototype for a computer, but soon discovered that in postwar England many bureaucratic restrictions had been restored, making work more difficult. During the war, German messages had to be deciphered as quickly as possible by whatever means, so scholars were given a free rein in choosing their methods of work. All that mattered were results. After the war, old restraints were reinforced. In addition, Turing did not really feel at home in Cambridge, so he accepted a position at the University of Manchester where researchers were also attempting to construct a computer.

During his stay in Manchester he published some more influential articles, among which was the well-known article on artificial intelligence in which he presented his famous test designed to determine if a computer could really think. According to the Turing test, as it was later called, a machine can be said to be capable of thought when, on the basis of its answers, no one can tell whether it is a human or a machine that is answering the questions.

On the 8th of June 1954 his cleaning lady found him dead in his bed. On his night table was a bitten apple poisoned with cyanide. Although there was no death note to be found it is almost certain that he had committed suicide. His mother firmly believed that her son's death was the result of his carelessness after a chemistry experiment, but, as one of his biographers argued, it is more likely that Turing took his life in a way that would allow his mother to give the act an interpretation she could live with.

Many years after Turing's death, a rumor spread that the bitten apple, which is today the trademark of the Apple company, was in fact a tribute to the tragic life story of "the father" of computers. This has never been officially confirmed.

Chaos and the butterfly effect

In the 1960s there were very few computers available for research purposes. One of the lucky few who had the opportunity to use a computer in his work was the American meteorologist and mathematician Edward Norton Lorenz from Boston's MIT. He set himself the task of creating a simple weather simulation. He simplified the physical equations that described the movement of air currents, so he could enter them into what was, from today's viewpoint, a very slow computer. The machine then used the starting conditions to calculate how the weather within the model would change.

A computer simulation of weather

In the winter of 1961 Lorenz started wondering how such a model would function over longer periods of time. Lorenz's computer had already completed a simulation for a short span covering a few days. In planning his longer simulation, he decided to save time by commencing from the point where the computer's earlier predictions left off, rather than starting the experiment again from scratch. This way he would cut down the wait his laborious computer would demand. And so, after entering this preliminary data, which fast-forwarded his work to the middle of the longer period he wished to investigate, he set the program running and went out for a cup of coffee.

When he returned he took a look at the printouts and noticed that, for a while, the numbers followed the previous day's results closely, but then strange things started to happen. The similarities were increasingly disappearing until the simulated weather became completely different from the weather predicted by the same program one day earlier.

At first, Lorenz thought there must have been a power outage which caused the program to start working differently, but it turned out that the power was never cut. In his book *Chaos,* James Gleick describes the moment when Lorenz realized the true cause of the deviations within the simulation results:

> He suddenly realized. This was no malfunction. The problem was in the numbers he had entered. The computer processor had saved six decimal numbers: 0.506127. To save paper, the printouts only contained three decimals: 0.506. Lorenz had entered the shorter, rounded up numbers without thinking that the difference in the fourth decimal might be important.

That is when Lorenz stumbled across an important new finding, known today as "the butterfly effect".

In the beginning, Lorenz was convinced that the Earth's atmosphere was a perfectly ordinary system that could be described in detail as long as the appropriate physical equations were available along with sufficiently precise measurements of its current state. He was wrong. Even though the system of equations he was using could accurately describe the movement of air currents, he could not use them simultaneously to make precise predictions of the weather for the more distant future.

Using computer simulation he came to understand that even the slightest change in air movement at a certain point in time could cause a substantial difference in the weather forecast several weeks later. During a lecture in 1972, he summarized the phenomenon quite vividly with the question: Does the flap of a butterfly's wings in Brazil set off a Tornado in Texas? Since then the power of the most minimal factors to bring about much greater changes in a large-scale system has become popularly known as the butterfly effect, and the field of science examining such systems as chaos theory.

It did not take long before the butterfly effect was noticed in many other fields of scientific research. A similar susceptibility to small changes in starting conditions was discovered, for example, in an equation used to describe population growth in individual animal species through time. Slight differences in the parameters could completely alter the predictions. Similar chaotic tendencies were found in many other equations which at first seemed very simple.

The Universe as a clockwork mechanism

Throughout modern history the Universe was thought to function as a colossal machine which, in essence, was no different than any other mechanism. The generally accepted idea was that of the world as a gigantic clockwork system which started running at the beginning of time and had been independent ever since, moving according to the laws that had been determined at the very start. It was believed to be comparable to an ideal clock that needs no setting or winding, but invariably shows the correct time.

From this perspective, scientific work is a sort of reverse engineering: the end product is the basis for determining the instructions according to which it had been put together or the way it functions. Much as they might take apart a brand new cell-phone to find out how it is made, scientists attempt to figure out the laws that govern the world. The rules according to which the world functions are mathematical laws like Newton's laws of motion and gravitation. If we knew all the constituent parts of the system along with the rules or laws that apply to them, we should be able to describe the system in detail, and precisely predict its future development.

If predictions were ever unreliable or wrong this was thought to mean that something must be wrong with the measurements or the established laws. One could have made a mistake in measuring the starting conditions or based one's calculations on the wrong principles. The analogy of the mechanical device is only flawed because there is something wrong with the mechanism itself: it might have been damaged or influenced by coincidental external factors we could not have been aware of. If predictions turned out to be false, this was not caused by the system, but by some unknown external disturbance which influenced the way it functioned. Yet in the second half of the twentieth century such ideas about nature were proven to be inadequate.

Chaos and complexity

In science, chaos is another name for the conclusion that a system that can be described in detail using mathematical equations is completely unpredictable even when it is not influenced by any unknown external forces. The idea that knowing the equations and the starting conditions is enough to precisely predict further development no longer applies to chaotic systems. Even a slight deviation at the outset can result in significant discrepancies in the predictions. Perfectly simple and clear equations which, in themselves, do not contain anything coincidental or wayward can nevertheless function in a completely unpredictable manner. That is to say, however plain and regular a set of conditions might seem at the outset, chaos dictates that they are still capable of producing effects that we could never expect. Future development cannot be predicted, even though the system is not influenced by any external factors or coincidences.

Of course, our intuition is reluctant to accept that a small error in setting the mechanism could cause a big difference in its future functioning. This would mean that if we were to set the hands of two clocks only slightly differently, after some time they would show completely different times. Even if the initial deviation is so small that it cannot even be noticed, in a chaotic mechanism, it will soon become evident. One of the characteristics of chaos is that seemingly chaotic behavior can also occur in deterministic systems that are protected from any external source of coincidence. This is why the behavior of such systems cannot be predicted in the long-term: such systems are extremely susceptible to small irregularities in their starting conditions.

The consequence of this discovery was not merely the realization that many things cannot be predicted, even if we are precisely familiar with their mechanisms. It also turned out that very complex or sophisticated phenomena can be the result of completely simple mechanisms or equations that describe them. What makes the world chaotic also makes it orderly. Order and chaos have more in common than it might have once seemed.

The big question ever since the days of the Ancient Greeks has been how ordered patterns in nature could have been created, if they had not been created according to a pre-established plan. It always seemed that complexity could not be derived from something simple, but only from something that is complex itself. Scientific determinism evolved together with the belief that complexity is something that is constant. The order found in nature supposedly could not have been created spontaneously on the basis of simple rules. However, chaos theory has shown that a few simple laws can also create the highly complex phenomena present in nature.

Atoms

In the beginning the Universe was created. This has made a lot of people very angry and has been widely regarded as a bad move.

Douglas Adams (*The Hitch Hikers Guide to the Galaxy*)

Searching for the beginning of time

The Saturday of October 22 in the year 4004 BC was a very special day. For quite a while it was generally believed that it was on this very day at 6 o'clock in the evening that God had created the world. This exact day was determined in the middle of the seventeenth century after years of research by an Irish bishop called James Ussher. The date became so widely acknowledged that, at the time, it was even printed in editions of the Holy Bible.

Ussher took on the project of dating the beginning of time when he began writing a detailed history from the moments following the creation to the year 70 AD. It took him more than twenty years to finish his work, researching ancient sources and manuscripts in different languages. He tried to place these different sources, all of which had their own method of counting time, on a common time scale.

He was so thorough in his work that he sent his assistants to the far corners of the earth to search for old books and manuscripts which would help him fill in the missing parts in the chronology he was creating. He also used astronomical charts to reconstruct the dates of important events such as solar eclipses, and compared his findings to the records given in ancient chronicles. Even though his original endeavor consisted of several thousand pages of dense text, he is remembered in history for the single sentence in his mighty tome in which he determined the date and hour of the beginning of time.

The success of Ussher's calculations was largely due to the general belief, which had held for many centuries, that the world was no older than a few thousand years. All that had to be done to determine the time of the creation of the world was to analyze ancient sources and work back by counting the number of years that passed from, for example, the birth of some biblical hero or king to the moment his son or daughter was born. It was slightly unusual that some of Adam's descendants had their first children when they were well over a hundred years old, but the experts simply attributed this oddity to the favorable climate during that golden era, when there was an abundance of healthy food which enabled people to live longer.

Could the world be older than a few thousand years?

One of the first to determine that the Earth was substantially older than a few thousand years on the basis of accurate scientific experiments was a French scholar called Georges-Louis Leclerc, the Count of Buffon. He was born at the beginning of the eighteenth century and died only a year before the beginning of the French Revolution. His life's work was also a written account of the history of the world, but unlike Ussher he was not interested in the ancient civilizations but in the natural world. He was still a young man when he became interested in the works of Isaac Newton, which he read and studied carefully in the original. His goal was to apply Newton's ideas to the natural history of the world, to develop a new theory of the creation of the Earth and attempt to put a date on the moment of its creation.

As a young man he traveled extensively across Europe. After his mother's death in 1732, he took over the family estate and bought a manor house which stood on a hill in his home town, Montbard. He knew how to invest his money and over time built a veritable scientific center in his home, with botanical gardens and an astronomical observatory in the manor's grounds. He also built an underground laboratory in which over the next fifteen years he conducted experiments to determine the age of the Earth.

Buffon was a very talented but also extremely hardworking scholar. He knew that success was above all things the result of hard work, so he instructed his servant to wake him up at five o'clock each day. Because he knew himself well enough to know that it would not be an easy task to get him out of bed this early he promised his servant a reward each time he succeeded. The servant was glad for the extra pay and developed a variety of tricks to wake his master up in the morning. Occasionally Buffon's sleep was so profound that it called for special measures. Once, when shouting, shaking and drumming proved ineffective, the servant resorted to pouring cold water. For this he did not get the sack, but received his usual bonus.

In 1734, Buffon's reputation as a scholar earned him a place in the Academy of Sciences, and five years later he became Keeper of the French King's Royal Gardens (*Jardin du Roi*, later *Jardin des Plantes*), a post he retained until the day he died. The Royal Gardens were not merely a botanical garden, but also one of the leading scientific institutions of the time.

The comet that created the Earth

The idea about the creation of the Earth which came to preoccupy Buffon for much of his adult his life came as a response to a theory of Newton's in the *Principia*. Buffon found it strange that all of the planets known at the time orbited the sun in the same direction and on approximately the same plane. The probability that such an alignment is merely a coincidence is very small. Buffon was unsatisfied with Newton's conviction that the planets move in this orderly way because they were set in motion by some intelligent creator of the solar system. Instead he wanted to find a natural cause which would explain the order of the planets.

Buffon's theory was that the planets were created as the result of a comet colliding with the sun. In the collision, a large fragment supposedly broke off from the sun and shattered, leaving pieces that later came together to form planets. This is how Buffon also explained the orderly movement of planets: they were set moving, he reasoned, by a single event which put them on their course around the sun at the same time and in the same direction.

According to Buffon, the Earth was created from a part of the sun, which meant that, at the beginning, it must have been as hot as the stars. On its journey away from the sun it naturally started to cool down, but the cooling period lasted for a long time. The theory enabled Buffon to calculate the approximate time at which the Earth was created. However, to make this calculation Buffon needed to know how fast scorching hot spheres of different materials would cool down and whether the rate at which they cooled depended on their size.

Buffon usually spent his winters in Paris where he was kept busy by administrative work for the Royal Gardens, and the summers on his estate where he dedicated most of his time to scientific work. There, on being forcibly woken at five, he normally spent an hour answering letters. At six he took a stroll through his gardens to the study where he wrote his books. The records kept by his contemporaries reveal that he was also a man of strange whims. He paid great attention to his personal appearance and supposedly needed the service of a hairdresser up to three times a day, and would spend his Sundays promenading through the town in the latest Paris fashion.

Cooling balls in the manor's basements

It took him a number of years to carry out the experiments which made him famous. Down in the basement of his manor In these he timed the rates at which balls of different sizes and materials cooled down. Strange rumors spread as to what was going on underground, since the Count often invited young girls to take part in his experiments. Buffon skillfully defended his methods to suspicious villagers by arguing that the girls' young and tender hands were essential to the success of the experiments since only they were capable of accurately sensing when the balls cooled down to the right temperature.

After years of experiments and calculations, Buffon reached the conclusion that an iron sphere the size of the Earth would require a little less than a hundred thousand years to cool down from the temperature of the sun to the present temperature. This was inconceivably long considering the ideas about the Earth's age which were held at the time to be true. Later, taking into consideration that the Earth was not composed solely of iron, he slightly adjusted his estimate, but it remained substantially larger than standard Christian interpretations could allow.

Buffon's writings soon reached the attention of theologians at the Sorbonne, who demanded that he take back his claims. He complied, and retracted his theory, simply because he did not want to waste time on unimportant disputes. In a written disavowal, he explained that he was withdrawing any of his claims about the age of the Earth which disagreed with the teachings of the Holy Bible. At the same time he made it self-evident that his retraction was not really sincere. However, it allowed him to keep his prestigious post as Keeper of the Royal Gardens and continue his research in peace.

In his history of the Earth Buffon argued that the planet cooled fastest in the parts closest to the globe's poles. This hypothesis, which placed the cradle of life on Earth in freezing Siberia, delighted the Russian Queen Catherine the Great, who sent Buffon a gift of expensive furs, jewelry and gems as a sign of gratitude.

Although in his published works Buffon estimated the age of the Earth at about a hundred thousand years, in his private notes he speculated that the world might be much older still. In the end, he surmised that the Earth could be several million years old. This was no more heretical than saying it was older than a few thousand years, as Ussher had asserted, but since he lacked enough solid evidence to support his claims he decided not to publish them. Although Buffon was a long way short of the present estimate of the

Earth's age, which is set at 4.5 billion years, by basing his calculations on experimentation he opened the way for an important era in the history of science. Geology was to become one of the principal scientific disciplines and lead to some of the most unsettling discoveries of the next century.

The priest who came up with the Big Bang

The Big Bang theory is one of the most popular scientific theories today. It states that the universe began as an extremely hot and dense substance which expanded and cooled with time until, 14 billion years later, it acquired the form that we observe today whenever we look at the sky on a clear night. The theory is well supported by experiments and represents a good example of the almost incredible capacity of modern science for gathering a small amount of data on our little planet and using it to reconstruct the history of the entire universe from its first moments on.

Despite its popularity not many people today are aware that the first to postulate the Big Bang theory was a Belgian priest called Georges Lemaître, who dedicated his spare time to science and was among other things a friend of Albert Einstein. Lemaître gave his vision of the Big Bang the slightly less catchy name of "the primeval atom theory". Einstein greeted the idea much more enthusiastically than his the wider scientific community would until the 1960s. For it was only then that two American scientists accidentally discovered that microwave radiation, which can only be explained by the conclusion that the universe was once much hotter than it is today, was reaching us from all the directions of the universe.

The universe as a radioactive atom

Of course, Georges Lemaître was not merely an amateur scientist. Even though he was ordained a Catholic priest in the fall of 1923, he always remained true to science. After earning his doctoral thesis in mathematics the same year he was ordained, he began his postdoctoral studies at Cambridge under the tutelage of Arthur Eddington, one of the most eminent astronomers of the time. It was with him that Lemaître became acquainted with the latest discoveries and the newest research methods in astronomy as well as cosmology, which was gradually becoming a branch of science within which experimentally verifiable theories could be formed.

After a year spent in England, he left for Harvard in the USA where he collaborated with Harlow Shapley, another great name in contemporary astronomy. At Boston he also studied for some time at MIT, where he became even more drawn to cosmology and started to examine the models of the universe implied by Einstein's general theory of relativity. He obtained a doctoral degree at MIT before returning to Belgium and becoming a professor at the Catholic University of Leuven.

Soon after his return, he published the first drafts of his theory of the origin of the universe. Even though one might have expected that, as a Catholic priest, he would look to the biblical story of creation when forming his idea about the Big Bang, this was far from the case. His inspiration came from neither astronomy nor relativity, but from a completely new and still more radical theory, quantum physics. More specifically, his idea of the primeval atom was inspired by work on the phenomenon of radioactivity.

Is the creation of the universe a mere coincidence?

It is well known that radioactive elements are not stable, but decay with time, releasing energy. Even though one cannot predict exactly when a particular radioactive atom will disintegrate completely, it is possible to predict with great precision when, for instance, this process of decay will break down one half of the radioactive atoms in a multitude of identical particles. On becoming acquainted with the principles of quantum physics, Lemaître asked himself whether it might be more than a mere coincidence that the half-life of some radioactive elements comes very close to estimations of the age of the universe, based on measurements of its rate of expansion. Could it be that we are living in some worn-out version of a radioactive universe?

According to Lemaître's original idea the universe began as a kind of a "large radioactive atom" which is why we can determine its exact age. The analogy between the universe and radioactivity was also of key importance in terms of overcoming the famous problem of explaining the very beginning of time. If we wanted to give a causal explanation of the first moment, we would have to refer to an even older event, but we cannot do so for the very reason that we need to explain the initial event first. Many different thinkers throughout history have dealt at length with this problem, but Immanuel Kant outlined it most famously as it as his first antinomy of pure reason.

The essential thing one should keep in mind when trying to understand radioactive atoms is that their disintegration is completely random. It is impossible to predict when exactly a particular atom will disintegrate, which is also why there no exact cause for its decay can be determined. It is completely unpredictable: we can estimate only the probability of decay occurring within a certain period of time.

Lemaître extended this idea about the coincidental decay of the radioactive atom to his idea of the universe as emerging from a primeval atom. He reasoned that this primary particle, like a radioactive atom, would have been governed by the laws of quantum physics. Even though it does not describe exactly the same types of event, the analogy that Lemaître tried to establish is quite obvious. Just as a particular example of radioactive decay has no cause, even though it is dictated by exact laws of quantum physics, the same might be true of the universe, which still does not mean that exact physical laws do not apply to it.

The universe began shortly before the beginning of time

In 1931 Lemaître described his idea in an article published in *Nature* magazine, in which he wrote:

> If the world began with a single quantum, the notions of space and time would altogether fail to have any meaning at the beginning; they would only begin to have a sensible meaning when the original quantum had been divided into a sufficient number of quanta. If this suggestion is correct, the beginning of the world happened a little before the beginning of space and time.[1]

At first glance, this way of thinking might be something that one would never expect from a Catholic priest, but anyone who is familiar with at least some Christian theological thought will quickly see how this theory leaves the door open for the doctrine of creation. Just as God, according to this doctrinc, is absolutely free in his creation of the world and is not subjected to any kind of limitations, thoughts or higher purposes, Lemaître's primeval atom is completely free in terms of the moment it "chooses" to disintegrate.

Although the analogy with radioactive decay was initially of key importance, it became less and less relevant the more the theory grew. Lemaître himself developed his theory in the years to follow, re-founding it on the general theory of relativity, but it was considered much too eccentric at the time to be accepted within the wider scientific community as a serious description of the actual universe. In science, a rule exists that the more unusual a theory is, the stronger the arguments and experimental proofs it needs in order to be accepted. As mentioned above, it was the discovery of cosmic microwave background radiation which led to scientists taking Big Bang theory seriously. But this only occurred two decades after the end of the Second World War, by which time the idea of the primeval atom had long been regarded as little more than an exotic hypothesis. Lemaître died soon after

---

[1] G. Lemaître, "The Beginning of the World from the Point of View of Quantum Theory.," *Nature* 127, no. 3210 (5, 1931): 706-706.

his visionary idea about the origins of the universe was largely confirmed by the discovery of cosmic microwave radiation.

In 1936 he was elected a member of the Pontifical Academy of Sciences which advises the Pope on matters relating to science. In 1960 he became its president, a post he held until his death in 1966. He always defended the belief that there is no ideological conflict between science and religion. He claimed this to be true as a high dignitary of the Church as well as one of the most esteemed scientists of his age.

He was not pleased when Pope Pius XII referred to science, including his particular area of expertise, to try and prove that scientific and divine truths were actually one and the same. Even today, high Church dignitaries often declare that science is coming to similar conclusions to those described in Church doctrines. According to them, The Big Bang theory describes the development of the universe all the way back to the beginning of time, with the exception of the first moment which is of course said to be the work of God

Lemaître strongly disagreed with such interpretations, which were subsequently defended by later Popes:

> As far as I see, such a theory [of the primeval atom] remains entirely outside any metaphysical or religious question. It leaves the materialist free to deny any transcendental Being. /.../ It is consonant with the wording of Isaiah speaking of the 'Hidden God' hidden even in the beginning of the universe.[2]

---

[2] 'The Primeval atom Hypothesis and the Problem of Clusters of Galaxies', in R. Stoops (ed.), *La Structure et l'Evolution de l'Univers* (1958), 1-32. Trans. Helge Kragh, *Cosmology and Controversy: The Historical Development of Two Theories of the Universe* (1996), 60.

How black holes are born

During the Cold War, both of the great political blocs sought ever more inventive means of keeping watch on the other. It was of particular strategic importance to detect immediately any movement of nuclear weapons, anywhere in the world. The detonation of a nuclear bomb on the Earth's surface could be detected quite easily on seismographs, but it was much harder if an explosion was carried out under conditions away from the ground where no sizeable tremor would be made. In the 1950s the Americans therefore began work on a top-secret project, codenamed Vela, to develop satellites capable of detecting nuclear explosions anywhere in the atmosphere or elsewhere in the vicinity the planet.

Could the Russians be testing their bombs on the moon?

At the time, widespread paranoia held sway: not knowing exactly what their Soviet rivals were capable of, the Americans tried making the detectors sensitive enough to spot a nuclear explosion even if it took place the distant moon. Incredible though it might seem, they seriously considered scenarios in which the Russians were capable of testing their nuclear weapons on the dark side of the moon.

In 1963, when the two great blocs finally signed a treaty banning nuclear weapon tests in the atmosphere, in outer space and underwater, in which they were later joined by the majority of other countries, it then became crucially important who had control over implementing the ban and monitoring both sides' observance of the treaty.

The American project Vela saw the launch of more than ten satellites, each equipped with sensors s capable of recognizing the brief pulses of extremely intense light released by nuclear explosions. Americans sent the first pair of Vela satellites into orbit only a few days after the treaty was signed. Predictably, the technology they carried was constantly being improved, so that in 1967 a pair of second-generation Vela satellites, fitted with still subtler detectors, was blasted into orbit.

Were they detecting explosions from a "galactic war"?

Soon after these new satellites were launched something completely unexpected happened. On July 2, 1967, they detected bursts of gamma rays which were different from those emitted by a nuclear bomb. Since the project was being carried out in complete secrecy, any extensive debate was impossible among astronomers at the time, but many great minds started to ask themselves what these radiation bursts from outer space could mean.

At the time, nobody was familiar with a natural phenomenon that could cause these flashes of such high-energy rays, allowing scientists with clearances to let their imaginations run wild. Someone suggested that the bursts occurred when comets made of matter and antimatter collided. Others reasoned that it was caused by small black holes exploding, or even that the satellites had discovered a galactic war between alien civilizations.

Over the next years, ten to twenty such bursts were observed, and later, when even more sensitive detectors were more developed, several hundred more. It was quite some time before scientists were able to examine gamma-ray bursts in detail and demonstrate that they came from the depths of outer space and not from a nearby source. The first article on the existence of gamma-ray bursts was not published until 1973, six long years after they were first discovered.

Where do the bursts come from?

As the question of what was causing these mysterious flashes was too difficult to answer at this stage, astronomers decided to establish where in the universe they were coming from. If they came from our galaxy they would be observable somewhere in the Milky Way, the familiar swirl of stars to be seen covering much of the sky on clear nights.

In fact, only a few people thought that the origins of these flashes could lie outside our galaxy in the first place. If they did, it was thought, they would simply be too powerful. Energy of such magnitude could not have been the result of any known process; if it existed, such a process would have violated many of the basic laws of physics, among others Einstein's famous $E=mc^2$ equation. The first credible theory hypothesized that the bursts are caused by small neutron stars from our galaxy when they are hit by an asteroid or a comet.

However, more thorough research into the location of these bursts revealed that they were equally spread across the entire sky and thus could not have come from our galaxy, but from the more distant universe. Of course, this only complicated matters further.

The next important project scientists took on was to determine the precise distance between the sources of these bursts and the Earth. The only way to measure this distance was by using a technique called redshift. We all know that the sound of an ambulance siren is higher when the vehicle is approaching us than when it is moving away. A similar phenomenon occurs with light in space. If a star or galaxy is moving away faster than another, it appears to be "redder" than the one that is moving away more slowly.

But there is another universal rule at work in outer space, called Hubble's law: the more distant galaxies move away faster than the closer ones. The relation between the two quantities is actually very easy to establish: the farther away a galaxy is, the faster it moves away. If we can determine the degree of redshift a galaxy displays, we can evaluate how far away in space it is using Hubble's law.

The brightest explosions in the universe

On May 9, 1997, astronomers observing a gamma-ray burst were able for the first time ever to determine which galaxy it came from. Such observations are logistically very demanding, because telescopes around the globe have to be coordinated to the satellites detecting the flashes, and react quickly enough to catch the remains of the light given off by the explosion.

When it detects a gamma-ray burst, a satellite sends an SMS to the astronomer on duty, who immediately notifies the large robotic telescope in the most appropriate location to trace the burst. The most suitable telescope, be it located on the Canary Islands, Hawaii or in Australia, immediately stops its current task and turns toward its new target in a matter of seconds.

After a number of such readings were successfully taken, it transpired that the sources of the bursts are situated in very distant galaxies. This means that the explosions generating gamma ray-bursts are definitely the most powerful that the universe has experienced since the Big Bang. Astronomers have gone some way to explaining such extreme events. In fact, these bursts are very narrowly directed rays which appear to be so powerful because they do not travel to all sides of the universe. They can only be seen when they travel in the direction of our galaxy.

The birth of black holes

At present, there are two possible models that can explain the formation of such high-energy light rays. According to the first model, two neutron stars collide and form a black hole. The second hypothesis postulates the idea of a very powerful supernova explosion, called a hypernova, which also leads to the formation of a black hole.

Hypernovas are supposedly generated by very massive stars, so large that their mass causes the fuel which enables stars to shine to combust very rapidly. In a hypernova, fuel burns out so quickly that stars cannot even leave the areas of the universe where they are born. And it is this rapid consumption of fuel that causes a massive star to shrink quickly which, in turn, triggers an explosion of enormous proportions in which a sheaf of gamma rays is released and a black hole created.

The calculations made in support of the two theories have produced models which are slightly at variance: and interestingly, astronomers have detected gamma-ray bursts corresponding to the first as well as the second scenario.

The origins of continents and oceans

Revolutionary scientific theories that turn upside down established explanations of natural phenomena usually require firm arguments to support them, before the scientific community starts taking them seriously. Even after clear evidence has been provided, quite some time is needed for a hypothesis to be accepted by experts in the field.

The theory that the continents drifted along the Earth's crust and collided with one another, resulting in the creation of mountain ranges, volcanoes and earthquakes, was not taken seriously for a long time because there was supposedly insufficient proof to support it. Today the theory of plate tectonics is considered one of the most important advances in the field of geology. During his lifetime, however, the man behind this mind-bending idea, which constituted perhaps the longest single leap forward in the history of earth sciences, was widely regarded more as a madman than a scientist.

Arctic enthusiast responsible for the geological revolution

Alfred Lothar earned a PhD in astronomy from the University of Berlin in 1904, but had always been more fascinated with the Earth than with the sky and the stars. He was especially interested in geophysics and in the quickly developing fields of meteorology and climatology. Even though today he is not particularly well-known for his achievements in meteorology, he contributed many original ideas and methods to this subject, and wrote a textbook which in the German-speaking area has long been recognized as the fundamental work in the field. Wegener had always been fascinated with Greenland, which he visited regularly on arduous and extensive scientific expeditions. There, he used weather balloons to follow the movement of air masses in the North.

One of the reasons Wegener loved Greenland was that he hoped its uncompromising landscape would confirm his unusual hypothesis that the continents were slowly but constantly moving. In a letter to his future wife in 1910, he wrote: "Is it not true that the eastern shore of South America and the western shore of Africa make a perfect match, as if the two were once one? I must examine this idea more closely." Since then he constantly searched for arguments that could best support his strange theory.

In the autumn of 1911, while rummaging through the library of the University of Marburg, he stumbled upon a scientific article about fossils that geologists had found in separate corners of the world, divided by vast oceans, but which were despite that fact identical. Traditionally, such correspondences were held to indicate the existence of land bridges which once enabled animals to cross the seas, but were later destroyed. However, Wegener had his own explanation for the matching specimens: the fossils were the same because the distant places on separate continents where geologists collected them were once close together.

The way the shapes of remote coastlines fitted together, along with the matches between far-removed fossils, convinced Wegener that the continents once constituted a single landmass. This super-sized continent in time became divided into parts, and these were now moving in all directions. Europe is shifting away from America, just as other continents are sliding away from one another. Wegener later named this vast "primordial continent" Pangaea. Another piece of evidence supporting the theory was the match between far-flung mountain ranges which, if put together, would run continuously together across the united continents. The perfect fit discovered between many mining provided Wegener's ideas with additional support.

Wegener set forward his ideas about continents drifting across the surface of the Earth in his book *The Origins of Continents and*

Oceans, published in 1915. He progressively revised and updated the book, completing new editions before his death. Unfortunately, the experts of the time had difficulties accepting that the ground was moving. The main problem was that Wegener could not explain the mechanism which either pushed or dragged the continents across the globe.

As an analogy, he often spoke of the process in terms of an icebreaker journeying through a frozen sea, and attributed the forces propelling the continents to the rotation of the Earth, but he failed to convince his fellow authorities. One geologist even calculated that a centrifugal force large enough to move continents would cause the Earth to stop rotating in less than one year. Another difficulty was that Wegener had false data to begin with, so he obtained some unusual results, like one which implied that Europe and North America were moving apart at 250 centimeters per year which was a hundred times faster than exact measurements showed many years later.

Death in Greenland

During the First World War, Wegener was enlisted as a soldier and served for some time on the Front. As a result of minor wounds was later transferred to the army's weather service. After the war, he returned to his post at the University of Marburg, but grew disillusioned there with the prospects of advancing his career. In 1924 he therefore accepted a position as professor of meteorology and geophysics at the University of Graz. In 1930, he embarked on another expedition to Greenland; a mission from which he would, unfortunately, never return.

Despite the severe cold, he set out with a friend to find a group of colleagues who were camped in the middle of the Greenland icecap and were running out of food. They managed to reach the camp and bring fresh supplies. After a couple of days' rest, they decided to return to base. The last time the rescuers were seen alive was on November 1, when they left the camp. Wegener's body was discovered by another expedition the following May. He was lying on reindeer skin, wrapped up in his sleeping bag. His piercing blue eyes were wide-open and, as the expedition leader wrote down in his journal, he seemed to be smiling.

He died only a day or two after his fiftieth birthday. The cause of his death was most probably a heart attack. His companion tried to reach the shore alone, but soon all trace of him disappeared. The expedition that found Wegener's body put up a big cross to mark the spot, but the movement of the Greenland ice soon knocked it down and it was not to be found in fifty years. Along with the cross, the ice had also swallowed Wegener's body.

Evidence supplied by military projects

During the Second World War and after, navies began using sonar to survey the depths of the oceans. Before, scientists had almost no knowledge of the ocean floors. The picture which sonar provided of the seabed by the sonar revealed a hidden world. Firstly surveyors were surprised to discover that there was a vast underwater mountain range, subsequently named the mid-oceanic ridge, in approximately the middle of the Atlantic Ocean. This range runs deep beneath the sea except in the north Atlantic where it rises above the surface of the sea as the piece of land we call Iceland. It was soon discovered that similar mountain ranges ran under other oceans, and that these mountain ranges were extremely volcanic. A process that occurs at mid-ocean ridges, where the ocean floor is extended when two tectonic plates move apart and new oceanic crust is formed through volcanic activity, is seafloor spreading.

Another important piece of evidence pointing back to the theory of moving continents also emerged from a military project. It came as another result of the superpowers' constant surveillance of each other's nuclear testing programs. As mentioned earlier, because an atomic explosion has the seismic power of a minor earthquake, it can be observed on seismograph readouts. With this in mind, the Americans installed several hundred devices for measuring seismic waves and used computers to analyze the data in a military base in Colorado. After a detailed examination of the measurements, they were able to determine where under the sea floors many of the earthquakes began, and at the same time discovered that seismic waves travel more slowly below the upper layer of the lithosphere than they do through solid rocks. Under the upper lithosphere, at depths of between one hundred and two hundred kilometers beneath the surface, lies the plastic asthenosphere over which the solid plates carrying the continents move.

On the basis of these and several other discoveries, the hypothesis of the drifting continents was revived. By the end of the 1960s it had been generally accepted, although only after a couple essential improvements had been made to Wegener's original idea. Continents do not plough through the bottom of oceans like an icebreaker through a frozen sea, but form vast tectonic plates with the ocean floors. It is not only the continents that move, as Wegener thought; the ocean floor moves as well. Continents on separate tectonic plates are like loads on conveyor belts moving in different directions. The face of the Earth consists of ten large plates which move and are pushed one over the other. When two "loads" moving on two separate plates collide, a mountain range rises. Similarly, the movement of continents explains other features of the Earth's surface. Today, the theory of plate tectonics is generally viewed as one of the essential theories on the structure of our planet.

In search of the perfect machine

By the first half of the nineteenth century the world was already in the midst of the industrial era. There were a growing number of factories looking to produce as much as possible at a low cost. Looking to increase productivity, factory owners soon began asking themselves if there were any natural limits to the quantity of work that could be done in given circumstances. When we work too hard, people get tired and have to rest, otherwise they become ill. What about machines?

How does a steam engine work?

One of the first researchers who devoted himself to studying the principles of steam-powered engines was the French physicist and military engineer Sadi Carnot (1796-1832). Despite his early death from cholera and the fact that most of his notes were burnt due to the nature of his disease, a substantial number of his findings were preserved, many of which are still seen as fundamental to engine research. Through his study of steam power Carnot sought to find by which heat engines worked.

A heat engine, in general terms, is a device that converts thermal energy into mechanical output. The first heat engines, which caused a true technological revolution when discovered, functioned on the basis of the fact that water increases in volume when it turns into gas at boiling point. Of course, this is nothing extraordinary and can be witnessed every day in the kitchen, when the lid of a pot full of boiling water starts to rise. The mechanics of a steam engine convert the force that causes the lid to rise during cooking and use it, for example, to spin a wheel. That is how a steam engine can propel a locomotive or do any other kind of mechanical work.

Carnot tried to explain how steam engines worked by using the analogy of a water mill. In this analogy, water took the place of the hypothetical substance that carried heat, which engineers of the day termed a caloric. Nobody could say for sure what this elusive element or medium might be: but like the water in a mill a caloric was believed to run through the components of a steam engine, and the result of its flow was the mechanical work achieved.

Physics and Romanticism

At the beginning of the nineteenth century engineers also developed an interest in understanding and then harnessing other natural forces. Research in the fields of electricity and magnetism demonstrated that caloric was not the only agent in nature that could be gathered to perform work. Such inquiries raised the important question of how one kind of force could lead to the creation of another kind of force.

It is noteworthy that the search for a link between the effects of different natural forces was greatly influenced by the branch of Romantic philosophy which believed in a deep connection and unity between all the elements and forces of nature. Thinkers in Romantic circles were firmly convinced that the different forces of nature were somehow interconnected, and that this central bond could be found and observed.

One of the first to attempt describing the universal transformation of one form of energy into another was Julius Robert von Mayer (1814-78), who traveled in the tropics places as a ship's surgeon. Once when several members of the ship's crew fell together, Mayer took blood from each of them, following the age-old (and even by that time still standard) doctrine of blood-letting, to restore the balance of fluids in their bodies. Although a perfectly common practice, Mayer was surprised to notice how venous blood in a tropical climate was bright red in color, as arterial blood invariably was. This phenomenon had probably already been noticed by other ships' surgeons before him, but nobody had paid much attention to it. Mayer was the first to ask himself as a scientist why venous blood should be so bright in those conditions. He soon came up with a convincing answer: because in warm places the human body does not consume as much oxygen to maintain a constant body temperature as it does in colder places, more oxygen remains in venous blood, giving it a brighter color.

On the basis of this discovery he made the hypothesis that in the human body as well as elsewhere in nature different energies converted from one form into another while the total amount of energy remained unchanged. As a surgeon, Mayer had no formal education in physics or mathematics, so he could not able to describe his findings in the academic form that scientific circles would take heed of and accept. Because of this disadvantage no one took him seriously at first. It was only at the end of his life when he was already close to giving up that his essays were noticed by more renowned physicists who confirmed his contribution to knowledge.

God and machines

In this period discussions about the notion of energy were also based on economic theory. At the forefront was a parallel between the efficiency of natural mechanisms and that of social mechanisms or the developing capitalist system. One of the principal goals of economics was to establish how a system might be made as efficient as possible. So it was no surprise that engineers, thinking along similar lines, tried optimizing the work that nature could be made to do for people.

In addition to the economic incentives for refining natural and artificial machines, theological ideas were also significant. A great number of intellectuals believed that God had created nature to function as perfectly as possible. Searching for the perfect machine meant at the same time drawing nearer to God and the perfection he had infused into nature itself.

The outline of the law governing the conservation of energy (known now as the first law of thermodynamics) proposed by the researcher of machine energy James Prescott Joule (1818-89) stems from the belief that what God has created cannot be destroyed or eradicated by a natural process. This is what he wrote: "Natural phenomena, be they mechanical, chemical or biological are based nearly entirely on the continuous conversion between at-a-distance forces, life forces and heat."

Joule was the son of a wealthy brewery owner, so he had no financial worries and could devote his time to experiments in the field of mechanical engineering. Arguably his most important contribution to the science of energy was the exact measurement of the mechanical equivalent of heat. Joule was the first to measure exactly how much a certain amount of water heats up if we stir it or perform some other type of mechanical work on it. This approach laid the experimental groundwork for the equations which allow us to calculate how much energy may be derived from an input of energy in another form, or in specific cases how much mechanical work it may enable a machine to carry out. .

Even machines have their limitations

William Thompson (1824-1907), better known as Lord Kelvin, also viewed nature as a gigantic steam engine. He thought the search for the most efficient machine possible was also the search for the master principles of nature itself. He wanted to find out how natural processes worked, so that he could apply the same solutions to artificially created machines.

In a series of articles published in the mid-1800s, Thomson laid the foundations for modern thermodynamics, the technical term for the branch of science which explores the laws of heat engine functioning. He formulated the first two fundamental laws of thermodynamics: the law of energy conservation and the law of entropy. Thompson was also the first to start using the term energy in the accepted sense it has in modern physics. Previously energy could signify "force" or "power", words which also now have very specific applications for physicists.

His formulation of the law of entropy was very important because it influenced the perception of nature or the world as a whole. The essence of this law is that the quantity of work that can be done by nature is not infinite. During the conversion of heat into work all the heat can never be transformed into work; a part of it always remains in the form of heat. In an ideal heat engine the ratio between the emitted and received heat would equal the ratio between both absolute temperatures at which the process takes place.

To put it another way, Thomson's discovery means that in order for heat to be converted into work there is always a difference of temperature. In addition this means that the amount of heat available for work is thus decreasing. A further theoretical implication of this principle is that someday, when no difference of temperature is recorded – when it is impossible, that is, for a drop in temperature to yield any observable output of work – what is known as the *Heat Death of the Universe* could occur. The reserves of energy available in nature for performing mechanical work are thus not endless, and indeed are shrinking. The conclusion suggested by the law of entropy is pessimistic: the amount of natural energy useful to humanity is irreversibly diminishing with time.

How to release the energy of atoms?

Leo Szilard was sitting in the lobby of his hotel in London reading the *Times*. It was Tuesday, September 12, 1933. In that issue, the newspaper featured an in-depth report on a scientific conference where the renowned physicist Ernest Rutherford had lectured on the theories others had advanced of gathering and using the energy which was supposedly stored in atoms. Journalists gave a detailed account of Rutherford's lecture, one that annoyed Szilard immensely. Rutherford claimed that "anyone who expects to find a source of power by transforming atoms is talking moonshine", and most of the scientists of the time believed him. But Szilard thought otherwise.

Rutherford's statements on the impossibility of nuclear energy made Szilard so angry that he decided to dedicate himself to solving this problem and prove the famous physicist wrong. He left the hotel and took a walk through the city to clear his mind. Before he had strolled very far down one of the nearby streets an idea flashed before him. He had thought of a way to release the energy of an atom. "I stopped for a red light at the intersection of Southampton Row [at Russell Square]," he later recalled. In a few moments cars started to slow down and stop and pedestrians were able to cross the road. It was then that Szilard realized how it would be possible to release the enormous amounts of energy stored in every atom. He would have to create a chain reaction using neutrons.

"Eureka" at the pedestrian crossing

> As the light changed to green and I crossed the street, it suddenly occurred to me that if we could find an element which is split by neutrons and emit two neutrons when it absorbed one neutron, such an element, if assembled in sufficiently large mass, could sustain a nuclear chain reaction. I didn't see at the moment just how one would

> go about finding such an element, or what experiments would be needed, but the idea never left me. In certain circumstances it might be possible to set up a nuclear chain reaction, liberate energy on an industrial scale, and construct atomic bombs.

During the winter, Szilard put his ideas down on paper. The concept he was dealing with was completely new: the existence of the neutron particle had only been proven a year earlier. The crucial discovery was that neutrons, unlike protons with which they constitute the nucleus of an atom, are electrically neutral and are not influenced by the electrical repulsive force, so they could penetrate the nucleus from outside and reshape it in such a way that it would become unstable and split it into two parts.

Naturally, finding an element which would emit energy and free neutrons after being split up was essential. According to Szilard, this method of inducing what he called nuclear fission would result in the release of energy, stored up to that point in the atomic nucleus. When this nucleus is split apart by the neutrons, two new nuclei are created: because the combined energy of these new particles would, he argued, be smaller than that of the original nucleus, the excess energy would necessarily be released in the form of heat. If this moment of fission released further neutrons, which then collided in the same way with neighboring nuclei, a chain reaction would take place, releasing unimaginable amounts of energy.

The most important challenge in achieving this was to find an element which would, during fission, release energy as well as free neutrons. At first Szilard surmised that the appropriate elements were beryllium and iridium, but this supposition turned out to be wrong. Only later did it become evident that the most suitable element for releasing nuclear energy was uranium.

A writer's prediction of the future proves more accurate than a scientist's

To finance his research he first turned to a factory-owner Hugo Hirst. It is interesting that he did not enclose scientific articles with his letter of request, but a passage from the novel by H. G. Wells, *The World Set Free*. In the novel, a scientist called Holsten invents a way to release the energy stored in atoms. "Of course, all this is moonshine," Szilard wrote in his letter to the factory-owner, intentionally citing Rutherford's lecture, "but I have reason to believe that when it comes to seeing what industry might do with the latest discoveries in physics, writers of fiction may well be offering more accurate predictions than scientists." It was none other than H. G. Wells who invented the term "atom bomb".

Soon after this, Szilard faced an important decision. Should he reveal his findings on nuclear energy to the public by publishing them and risk his idea going into the wrong hands, or should he keep it a carefully protected secret? Holsten, the scientist from H. G. Wells's novel, decided that he was no more than an insignificant instrument in the grand machinery of progress and change. Even if he burned all his papers, someone else would come to the same conclusions in a few years. But Szilard decided otherwise. Instead of publishing his idea in a scientific journal, he decided to keep it secret.

First he thoroughly described his discovery of how the energy of atoms might be tapped and exploited. Then he patented his idea of critical mass and the chain reaction produced by means of neutrons. In 1935 he transferred the rights to the patent to the British Army on condition that it remained an absolute secret. During the following years leading up to the Second World War, all those with clearance to read Szilard's work did their utmost to prevent it reaching Hitler's scientists. Meanwhile in 1939 Szilard confirmed his theory in the laboratory. In February that year, an experiment was carried out at Columbia University, New York which later led to the first controlled chain reaction, and opened the era of nuclear energy. Szilard built the nuclear reactor in Manhattan with Enrico Fermi, an expert on bombarding atomic nuclei with neutrons. Several years earlier in Rome, Fermi had already discovered that decelerated neutrons fuse better with atomic nuclei than faster neutrons, so it was important to find a substance that could slow neutrons down. One suitable option was heavy water (water highly enriched in the hydrogen isotope deuterium), but this was difficult to obtain in sufficient quantities. To perform the experiments a more accessible alternative had to be found.

How to beat the Nazis?

As Szilard feared, German scientists were also thinking of ways to build a nuclear reactor. However, they had the same difficulties obtaining sufficient heavy water, forcing them also to search for another solution. Both teams tested graphite, but fortunately the Germans gave up on this possibility after concluding, too hastily, that graphite absorbed too many neutrons. Szilard, on the other hand, realized that the problem lay not with the graphite itself, but the quantities of boron added to it in the manufacturing process. Pure graphite in fact proved highly suitable as a "brake" for slowing down neutrons in a nuclear reactor. When this became apparent, Szilard was much more aware of the gravity of the moment than Fermi. In the summer of 1939, Fermi left for the University of Michigan to study cosmic radiation, while Szilard decided to turn to his old colleague Albert Einstein who, at the time, lived on Long Island, a hundred or so kilometers outside New York. Szilard presented his idea, which was actually based on Einstein's famous and revolutionary equation $E=mc^2$. Einstein responded by saying that he had never thought of anything like this when he thought of how his proof relating mass to energy might be applied.

On his second visit to Einstein, Szilard brought with him a letter he had written to President Roosevelt warning him of the danger that Hitler's Germany might also be trying to create an atomic weapon from nuclear fission, and suggesting that the U.S should seriously consider developing the technology itself. It is reported that Einstein read and signed the letter, which later turned out to be extremely important, wearing only his dressing gown. The letter began with the words "Some recent work by E. Fermi and L. Szilard..." and was dated August 2nd 1939.

Of course, Szilard did not simply send the letter to the President by mail as it might have come into the wrong hands, but instead asked an acquaintance in the administration, one of the President's economic advisers, to take it directly to the White House. When Roosevelt finally read the letter he realized that something absolutely must be done "to see that the Nazis don't blow us up". He founded the Uranium Committee, of which Szilard was made a member, and which set in motion the research that ultimately led to the creation of the atom bomb. It also brought about the nuclear reactors which we use to produce great amounts of electrical energy today, and which many environmentalists still view as one of the most acceptable alternative energy sources in the ongoing fight against global climate change.

What are quantum particles telling us?

Asking oneself which is the greatest discovery or the most important invention of the last century might at first seem naïve. Scientific and technological advances have been made at such a rate in the last hundred years that the choice seems nigh on absurd. However, it is sometimes a simple question like this - often posed by a well-intentioned journalist - which forces experts to think about things they would normally not take time to consider.

The discovery of randomness

The choice of the most important discovery of the past century, which was published in *Nature* by Anton Zeilinger, an Austrian physicist, will surely come as a surprise. This is what he wrote:

> The discovery that individual events are irreducibly random is probably one of the most significant findings of the twentieth century. Before this, one could find comfort in the assumption that random events only seem random because of our ignorance. /.../ But for the individual event in quantum physics, not only do we not know the cause, there is no cause. /.../ There is nothing in the Universe that determines the way an individual event will happen.[3]

It is often said that no one understands quantum physics. At least that is a claim made by several distinguished physicists who have been awarded a Nobel Prize for their research on the quantum world. By this lack of understanding they normally refer to the unusual traits of quantum particles that are impossible to explain through any analogy with everyday life. Quantum particles are crazy, as Richard Feynman once remarked, but all to the same extent: all of them can at the same time travel along different

[3] Anton Zeilinger, "The message of the quantum," *Nature* 438, no. 7069 (December 8, 2005): 743.

paths, appear in different places, and possess incompatible characteristics; but that does not seem to disturb them at all.

The physicist that teleported particles

Anton Zeilinger is definitely one of today's most important quantum physicists. A few years ago he became famous for his successful attempts to teleport quantum particles, an achievement which impressed mainly science fiction fans, and also made its way onto the covers of the best-known science journals.

Anton Zeilinger spent a number of years working mostly on experimental quantum physics and studying quantum teleportation, quantum cryptography, quantum computers and interference experiments with multi-atom molecules, he has dedicated the last couple of years to writing about the wider and more popular interpretations of quantum physics, or in other words the problems associated with this simple question: what does all this quantum nonsense even mean?

In 1996 he began critically examining the ways quantum mechanics had been written about, concentrating on how the pioneers of modern physics had discussed their work in their private correspondence. He gathered his findings in a concise article which concluded that quantum physics needs a clearly formulated basic principle to sum up its essence.

The basic principle of the theory of relativity is, in very simplified terms, that nothing can travel faster than light, that the laws of nature are the same for any observer and that we cannot separate gravity from acceleration. According to Zeilinger, quantum physics needs something just as simple, clear and universal.

In 1999, after three years' further thinking, Zeilinger published an article entitled *A Foundational Principle for Quantum Mechanics*[4]

[4] Anton Zeilinger, "A foundational principle for quantum mechanics," *Foundations of Physics* 29, no. 4 (1999): 631–643.

in which he formulated the basic principle of quantum physics, using the principles of the theory of relativity as a model. His proposal for the basic principle of quantum physics is: The elementary system carries one bit of information.

The smallest carriers of data

Naturally, these questions instantly arise: what is information and what is the elementary system? As he says himself: information is nothing other than the answer to the questions asked. The bit is the smallest piece of information that can still mean something. It is the smallest, indivisible unit of information. It simply states whether a statement is true or false. One could also say that it is the answer to a question which can only have two possible answers: yes or no. One bit of information can be represented simply as the presence or absence of a signal: a light turned on or off, the magnetization on a tiny piece of a hard disc, an indentation on the surface of a CD.

When Zeilinger was thinking about information in the quantum universe he also asked himself the important question of the relation between the physical size of a system and the quantity of information that the system can carry. A system which is two times smaller than another will probably carry two times less information. And if we continue to divide a certain system by two we are bound to eventually reach a limit where our system can only carry a single piece of information, one bit. That is how Zeilinger defined the elementary system as the carrier of a single bit of information.

Bits and qubits

As more and more physicists became interested in information problems in the quantum universe, the term quantum bit or qubit started to replace Zeilinger's elementary information carrier. The qubit is thus the basic carrier of quantum information. It is in a way an atom in quantum terms. Using this new term we could reformulate Zeilinger's basic principle into something like: One qubit can carry one bit of information.

However, the problem with qubits is that it is not possible to double them. A qubit can not be cloned without destroying the original which we wish to double. A qubit can also never be read with complete accuracy. If we discovered a process to multiply it, we could use the many identical copies to examine it thoroughly and precisely define it. As it is impossible to double it, one measurement of a qubit only reveals a single bit of information and the essence of the quantum universe remains always, to a certain extent, random.

According to Zeilinger, all problems stem from the very fact that information is quantified. We simply can-not acquire less than one bit of information about the world. It is the absolute minimum which at the same time means that the resolution of the world itself is limited to one bit of information. One qubit only gives us one bit of information. One qubit can only answer a single yes or no question. If we continue to question it via further experiments, its answers will not make any sense at all or will be, as Zeilinger puts it, objectively random.

The message of quantum

The fundamental problem of quantum physics lies in the difference between the qubit and the bit. The qubit contains more regularities than we can see. Zeilinger says:

> So, what is the message of the quantum? I suggest we look at the situation from a new angle. We have learned in the history of physics that it is important not to make distinctions that have no basis – such as the pre-newtonian distinction between the laws on Earth and those that govern the motion of heavenly bodies. I suggest that in a similar way, the distinction between reality and our knowledge of reality,

> between reality and information, cannot be made. There is no way to refer to reality without using the information we have about it.[5]

Just as the special theory of relativity is based on the impossibility of differentiating between inert observers (the principle of relativity) and the general theory of relativity on the impossibility of differentiating between gravity and acceleration (the equivalence principle), quantum theory is supposed to be founded on the impossibility to differentiate between the real world and information about it: the laws of nature should not separate reality from information.

Objective randomness

In quantum physics, the problem is not that our capacities for understanding the diversity of the quantum universe are too limited, but the fact that inevitable randomness is inherent to the very structure of the world and as such cannot be influenced.

Zeilinger calls this fact a state of objective randomness:

> But the individual measurement result remains objectively random because of the finiteness of information. I suggest that this randomness of the individual event is the strongest indication we have of a reality 'out there' existing independently of us. Maybe Einstein would have liked this idea after all.[6]

---

[5] Zeilinger, "The message of the quantum."

[6] Ibid.

What can we know about the world of atoms?

At the end of April 1926, the then only twenty-five-year-old Werner Heisenberg gave a lecture at the prestigious Physics Symposium at the University of Berlin. The title of his speech had attracted all the eminent German physicists of the time, among others the already very famous Albert Einstein. In front of this demanding audience, the young physicist presented a new theory on the behavior of atomic particles that he had recently developed in cooperation with other young colleagues. Even though everyone was interested to hear what theory these young scientists had come up with, the lecture turned out to be quite unusual.

At the time physicists were preoccupied with the problem of explaining why certain microscopic particles sometimes behaved like normal particles, and at other times as waves. This strange double nature of atomic particles posed a question to which nobody could provide a satisfactory answer. What was then unusual about Heisenberg's lecture was that the young physicist seemed to try very hard not even to mention what might be happening inside the atoms themselves. He limited himself strictly to demonstrating a mathematical theory which might be used to predict the results of experiments on such atoms, but without giving any explanation of what was actually happening to them. That was because the physicist himself did not know the answer to the burning question itself, but he had found a way to predict mathematically the atomic behavior which had raised the question in the first place.

After the lecture, Einstein himself came up to the young physicist and invited him to accompany him on his way home. Of course, Heisenberg was ecstatic about the offer. He was certain that Einstein would be very pleased with his approach to the problem of quantum physics, as this field of science had been named, because he had adopted a similar approach when working on his theory of relativity. Just as Heisenberg intentionally avoided delving into what actually happened inside the atoms and restricted himself to the characteristics of atomic activity that could be measured and calculated, Einstein, some decades earlier, when dealing with the question of time and space had systematically limited himself to individual readings and direct measurements, leaving to one side the deeper questions of what time and space actually were.

"It is the theory that determines what can be observed"

The innovative approach with which Einstein restricted himself to what could actually be measured, temporarily leaving other questions aside, enabled him to start a veritable revolution in physics and to redefine completely the discussion about the nature of time, space, energy and matter. Heisenberg hoped that, by applying this method to the world of quantum particles, he could make a similar breakthrough to that of his idol some two decades earlier.

Surprisingly, Heisenberg's idea of avoiding theoretical models and limiting his work to observable phenomena did not please Einstein at all. It is interesting that the older man's doubts did not stem from the concern that a purely mathematical theory was insufficient to explain what happened in nature. His reservations arose from his view that one should still try to apply available scientific theories when working from directly accessible physical data. This would include, for example, explaining the movement registered by a measuring device when an atomic particle collided into the instrument. Einstein claimed that theoretical frameworks were always essential to explaining empirical data, for the compelling reason that the data made no sense without them. For Einstein, that is, there was no such thing as the direct observation of the world independent of a theory to interpret it.

As Heisenberg later wrote down in his memoirs, one of things Einstein said was:

> But on principle, it is quite wrong to try founding a theory on observable magnitudes alone. In reality the very opposite happens. It is the theory which decides what we can observe. /.../ Only theory, that is, knowledge of natural laws, enables us to deduce the underlying phenomena from our sense impressions.[7]

Heisenberg defended his newly discovered theory of quantum mechanics, but he and Einstein quickly reached the conclusion that they still knew far too little about what went on in the world of atoms to reach any definitive conclusions. It was also apparent that, because of the great age difference – Einstein was almost twice his younger colleague's age – they were unable to have a completely relaxed discussion. In the following years it was Niels Bohr, an older Danish physicist closer to Einstein's age, who took over the debate on the important issues raised by quantum physics and also became one of the key scientists exploring the inner world of atoms. The discussion between Einstein and Bohr is regarded as one of the greatest intellectual debates of the twentieth century. It lasted almost thirty years, from the fifth Solvay Conference until Einstein's death in 1955.

Einstein would then become a little upset ...

In September 1927, at the Como Congress in Italy, Bohr presented his principal idea for explaining the world of atoms to his colleagues. Einstein was not present at this meeting, but he supposedly did not miss much, because Bohr's lecture, according to those who were there, was so condensed that nobody could understand it very well. Today, however, this very lecture is considered a milestone in the history of physics, because it established the central

---

[7] Werner Heisenberg, *Physics and Beyond: Encounters and Conversations* (Harper & Row, Publishers, 1972), 63.

idea upon which the interpretation of quantum mechanics is focused. It still has a place in many physics textbooks today.

According to Bohr, physics can only base its theories on what we can say about the world and not on what the world itself really is. While Heisenberg ignored the problem of how to imagine the world of atoms, Bohr took a step forward by showing that one can get a relatively good idea of what is going on by combining the two opposing models explaining the nature of atoms. Bohr's principal idea was that the world of atoms cannot be described using one consistent model only, but by applying several together.

Naturally, Einstein strongly disapproved of Bohr's idea, so he decided to put all his efforts into revealing a paradox between his theory of relativity and the claims made by the new quantum physics. This is how Heisenberg remembered the events that took place at the fifth Solvay Conference in Brussels in 1927:

> We all stayed at the same hotel, and the keenest arguments took place, not in the conference hall but during the hotel meals. /.../ The discussion usually started at breakfast, with Einstein serving us up with yet another imaginary experiment by which he thought he had definitively refuted the uncertainty principle. We would at once examine his fresh offering, and on the way to the conference hall, to which I generally accompanied Bohr and Einstein, we would clarify some of the points and discuss their relevance. Then, in the course of the day, we would have further discussions on the matter, and, as a rule, by suppertime we would have reached the point where Niels Bohr could prove to Einstein that even his latest experiment failed to shake the uncertainty principle. Einstein would look a bit worried, but by next morning he was ready with a new imaginary experiment more complicated than the last, and this time, so he avowed, bound to invalidate the uncertainty principle. This attempt would fare no better by evening, and after the same game had been continued for a few days, Einstein's friend Paul Ehrenfest, a physicist from Leyden in Holland, said: "Einstein, I am ashamed of you; you are arguing

against the new quantum theory just as your opponents argue about relativity theory." But even this friendly admonition went unheard.[8]

Can we learn more about atoms?

In 1930, Einstein arrived at the sixth Solvay Conference with an elaborate and carefully designed theoretical test which he had prepared in advance and was convinced could prove that something was wrong with quantum mechanics. However, after a detailed analysis, Bohr saw that this time Einstein had failed to take the effect of his own general theory of relativity into consideration, an omission which caused him to think he had found an error in quantum physics. If Einstein had taken the general theory of relativity into account, there would be no inconsistency.

After another failure Einstein no longer tried to prove that quantum mechanics itself was incorrect, but went at the problem from a different angle. He tried to prove that quantum theory was incomplete. In other words, he wanted to show that, in theory, it would be possible to say more about atoms themselves than quantum physics would allow. He wished to show that atomic particles had characteristics that cannot be explained by quantum physics, but could possibly fit into a different theory and could be measured as well.

He needed no less than seven years to find a way of proving theoretically that quantum physics was an incomplete theory which failed to account for all the characteristics of atomic particles. In 1935, he and his colleagues Boris Podolsky and Nathan Rosen at Princeton published a paper in which they described a theoretical experiment to prove that nature must know more about itself than what the equations of quantum physics can reveal.

When Bohr found out about Einstein's new paper, he immediately dropped everything he was doing and dedicated himself to

---

[8] Ibid., 80.

finding an error in Einstein's theoretical experiment. Three months later, an article with Bohr's answer to Einstein was already published. Nevertheless, Einstein did not agree with Bohr's defense of quantum physics and their debate came to a standstill. Both were certain that it was more or less a matter of a "philosophical" debate. A few decades later, though, another physicist, while reading Einstein's paper, realized how it would be possible to verify whether quantum physics was a complete theory or not. But that is another story.

MOLECULES

There is a theory which states that if ever anybody discovers exactly what the Universe is for and why it is here, it will instantly disappear and be replaced by something even more bizarre and inexplicable. There is another theory which states that this has already happened.

Douglas Adams (*The Hitch Hikers Guide to the Galaxy*)

Cannibals, insomnia and mad cow disease

Not long after the Second World War, Australian colonial officials assigned to the remote plains of Papua New Guinea noticed an unusual disease among the members of a small tribe called Fore. The locals called the ailment "kuru" which, in their language, means "to tremble". That is because those affected by the disease gradually became unable to control their muscles, started to shake and sometimes even laugh or cry uncontrollably. Once the first symptoms of the disease appeared, the sufferer's death would follow in a matter of months.

The unusual diet of the natives

In 1957 when Carleton Gajdusek, at the time a 33-year-old doctor, arrived to New Guinea, the stories about the mysterious disease, told by the colonial officers, immediately caught his attention. He set out to visit the tribe and began conducting a thorough analysis to discover the true cause of the illness. In order to research the disease Gajdusek had to gather samples of the tissue belonging to the diseased natives. Most of the time he managed this by buying a dead body and doing an impromptu autopsy, sometimes right on the kitchen table in his hut, and then saving the tissue samples in a shared refrigerator.

When, after carrying out different tests, he eliminated the possibility that the disease was being caused by viral or bacterial infections, he started examining all the environmental factors that could have contributed to the outbreak of the disease. He analyzed the food that the members of the tribe ate as well as the surroundings of their dwellings, but he was unable to discover the cause of the disease. Even worse: he failed to classify kuru by any of the main medical categories into which diseases are usually sorted. By all appearances, kuru was neither a genetic disease nor an infection, nor a psychosomatic disorder; and neither was it caused by the environment.

After long years of research, which he continued in the USA, Gajdusek succeeded in proving that the disease was transmitted through brain tissue. He used a sample of a kuru patient's brain tissue to infect the brains of a healthy monkey and it fell ill with the disease. In 1976 he received the Nobel Prize for Medicine for his research on this "completely new form of infection".

It soon became clear why the disease was spreading even more rapidly among the women and children of the tribe. It turned out that the kuru disease epidemic was caused by ritual cannibalism adopted by the tribe from their neighbors. It was typical of the tribe not to cook and eat their enemies, but their own people who had passed away. This culinary ritual represented a part of their custom of bidding farewell to the dead. Symbolically, the funeral was compared to digestion and the tribe shared different parts of the cooked corpse according to the hierarchical structure of their society; only feeding on the members of the inner family circle was forbidden. If a woman died, for example, her daughters-in-law would get her arms and legs while her sisters-in-law would get her buttocks and her intestines. If a man died, his testicles were given to their uncles' wives.

Cannibalism disappeared from New Guinea four decades ago after being successfully eradicated by the missionaries soon after their arrival. However, a few of the older female members of the Fore tribe still die of kuru each year, which indicates that the incubation period can last very long. According to evaluations, the disease has caused the death of almost 3000 natives.

Has the central dogma of molecular biology been violated?

However, Gajdusek is not the only researcher to have received the Nobel Prize for work in this interesting field of medicine. In 1997 the prestigious scientific award was also given to Stanley

Prusiner, who succeeded in substantially developing Gajdusek's findings and discovering the mechanism that explains how kuru and similar diseases actually work on a molecular level. (Gajdusek also became known for being one of the few Nobel laureates to have been sentenced to prison at an old age. In 1997, while living in the USA, he was sentenced to 18 months' imprisonment for child molestation. He found out about Prusiner's Nobel Prize while serving his sentence.)

In 1972 Prusiner, then a young neurologist based in San Francisco, lost a patient to Creutzfeldt-Jakob disease, which is well known today on account of the "Mad Cow" disease. As he did not know much about the disease at the time, he buried himself into literature and found out that it had been proved through tests on animals that Creutzfeldt-Jakob disease, along with scrapie (a disease of livestock) and the exotic kuru disease from Papua New Guinea, were all transmitted by the infection of healthy brain tissue with a sample taken from a diseased organ.

As the only research material he could get in sufficient amounts were infected sheep, he decided to try and isolate the agent that causes scrapie, a disease of the central nervous system which affects small cattle (sheep, goats and pigs). The method required for isolating such an agent was complicated because different types of molecules had to be extracted from the infected tissue. These molecules then had to be tested in order to determine whether they caused the disease or not. Or, to give a more detailed description of the process: a sample of brain tissue was taken from a diseased animal and put into a centrifuge in order to separate individual molecules. The further separated samples were then injected one by one into a mouse to determine which part was still infected. The process was then split into further phases to observe and determine which parts were still causing the disease.

The first hypotheses suggested the existence of so-called "slow viruses", but it turned out that the extract obtained from diseased brains could still cause an infection, even if all the nucleic acids which contain the viruses' genetic material were destroyed with radiation. It was also found, however, that by adding enzymes that deform proteins the sample did become less contagious, which meant that proteins were most probably the main cause of the disease.

Proteins are very important molecules in our bodies, often called nature's robots. They perform many kinds of different tasks in our cells. They are built from long chains of amino- acids which do not crawl around cells like long snakes, but form complex structures as soon as they come into existence. It is only in this form that they become active and capable of performing their various tasks. The protein called hemoglobin, for example, carries oxygen through blood.

Nobody was able to understand, though, how proteins multiplied in the absence of the DNA or RNA which contain the instructions for their production. According to the classical theory, the foundation of all modern biology, genetic material is stored in DNA molecules which provide the basic information cells need to produce the proteins which then perform different tasks in our bodies. According to this theory, also called "the central dogma" of molecular biology, the transfer of information is only possible from DNA to proteins and in no way in the opposite direction. A protein cannot create instructions for its own production and write them into the cell genome. It is rather always the cell that produces the proteins, keeping their blueprints stored inside its genetic code.

Bad molecules

Quite a few hypotheses attempting to solve the riddle of the mysterious infectious proteins were made, but in the end the most unusual one turned out to be the most accurate, and it was this theory which won Pruisner his Nobel Prize. As he was aware of the fact that publicity and promotion of a discovery is an important part of science, he needed an attractive name to suit his innovative idea. He wanted something simple, easy to remember, that would sound like already established scientific terms such as quark, proton, neutron, electron ... And so, the prion was born.

Usually, only one form of a protein is stable. Prions are special among proteins because they have two stable forms. The common, natural form can be found in healthy organisms while the other one causes the disease. The most extraordinary thing is that the "bad" prion has a great ability to recruit and convert the prions around it. If it meets a "good" prion it will most probably turn it into a "bad one" when they come into contact. The consequence of this conversion is that in the presence of merely a few "bad" prions the majority of the "good" prions eventually turn "bad". And this is exactly what happens in the case of prion diseases. Prions are, in fact, proteins that can multiply without the participation of genes.

In the case of a spontaneous occurrence of Creutzfeldt-Jakob disease, the first "bad" prions appear by coincidence and they immediately start their mission of conversion. In about a year after the first symptoms occur, the patient succumbs to the disease. Some families, fortunately there are no more than a hundred in the world, are genetically prone to suffer from this disease due to their slightly different genetic code. Nevertheless, others can be infected by "bad" prions as well. In some cases, people were infected while having their brains examined because doctors repeatedly used the same electrodes, transferring the diseased patient's "bad" prions to the healthy patient's brain, where the evil proteins set out on their deadly mission.

A mutation in the part of the genetic code responsible for the production of prion proteins can also cause an unusual disease, named fatal family insomnia. It was first noted in the 1980s in a respectable Italian family whose ancestors, upon reaching middle age, have for centuries died of unusual symptoms which were always accompanied by insomnia. Sometime between their 35th and 60th year of life some of the family's members found themselves suddenly unable to fall asleep normally. The problems could not even be remedied with potent sleeping pills and they continued to grow worse with every month until the patient, terminally exhausted, fell into a coma which was followed by death less than a year after the first symptoms occurred.

A cure for treating prion diseases has yet to be found. This is due to the fact that it is extremely difficult to prevent "bad" prions from leading other prions astray once they have entered a body. The usual kinds of drugs successful in fighting viruses and bacteria are powerless here. A prion inside a living body practically cannot be "killed".

EPO – The story about 2550 liters of powdered urine

Few people have probably ever heard of Eugene Goldwasser, a retired professor from the University of Chicago. This is not strange as he is neither a Nobel laureate nor an eminent figure in his field of science. However, his name definitely sounds more familiar if we mention that he dedicated several decades of his scientific career to finding the molecule of erythropoietin, commonly known as EPO.

Erythropoietin is a hormone that promotes the formation of red blood cells within an organism or, in simpler terms, provides cells with a better supply of oxygen. This, in turn, improves the endurance of athletes, making the hormone a popular prohibited stimulant which has recently also caused problems for the cycling champion Lance Armstrong.

In search of a cure against the effects of radioactivity

The story about Eugene Goldwasser, recently popularized by Merrill Goozner in his resounding book *The $800 Million Pill – The Truth Behind the Costs of New Drugs* (University of California Press, 2004), is interesting from several points of view.

On one hand, it tells the story of a typical scientist who was drawn so much to his research that he persisted in it despite decades of failed attempts and dead ends until all the long years of hard labor finally led him to an important scientific discovery. In the late 1970s, after decades of experimenting with sheep blood and after collecting and dehydrating 2550 liters of urine, he succeeded in producing 8 milligrams of pure natural human erythropoietin.

On the other hand, the book also tells the story of one of the first great commercial breakthroughs resulting from the biotechnological revolution in the field of drug production. Today, pharmaceutical companies produce synthetic EPO in large quantities and sell it for similarly large amounts of money. The biotechnology company Amgen, one of the main producers of EPO, creates more than a half of its multibillion dollar yearly profit on this hormone alone.

Unfortunately, Goldwasser does not receive his share of this enormous heap of money, founded on the production of "his" molecule, even though with more than twenty years of enduring work he made the largest contribution to isolating this molecule from the mass of others found in the human body. As he failed to patent his discovery at the time (he wrote to state financiers, asking them to take care of his patents, but received no response), he has to be content with the thirty thousand dollars graciously deposited by Amgen in the bank account of his laboratory at the University of Chicago each year.

The story about the discovery of erythropoietin, not unlike many other stories of science in the 20th century, begins during the Cold War, an extremely bountiful era for scientific research. Soon after the Second World War, the young Goldwasser was invited to join a research group commissioned by the state to discover ways of mitigating the consequences of a nuclear war. The government was interested in finding an antidote to the deadly effects of nuclear radiation on the human body. At the beginning of the 20th century, scientists had already realized that blood must contain a substance, which instructs bone marrow to produce the red blood cells that carry oxygen around the body. Scientists named this molecule erythropoietin after the term for the process of red blood cell formation called "erythropoiesis", even though they knew nothing about the molecule itself at the time.

Looking for a needle in the haystack

In 1955 Goldwasser was entrusted with a research mission to find the erythropoietin molecule and discover a way to produce it in large quantities, so that it could be used to cure radiation sickness. The search took longer than they expected: in fact it extended over more than two decades. However, if one takes a closer look at how demanding the job Goldwasser and his colleagues had taken on, the time spent to complete the task does not seem so long. A healthy person produces two to three million red blood cells every second, which adds up to nearly half a ton of blood in a lifetime, yet in the same period barely enough erythropoietin to make a tiny pill goes through the human body. The search for a molecule the scientists knew almost nothing about was therefore far from simple. It could be compared to looking for a lost coin on a long sandy beach.

During the first years of his research, Goldwasser wanted to find out which part of the body actually creates EPO. His assistants carefully removed organs from lab rats until they were certain that it was the kidneys that were responsible for the resulting bad blood. In the next phase they injected sheep waiting to be slaughtered with a chemical that destroyed all their red blood cells. They were convinced that the destruction of the red blood cells would result in a mass production of EPO which could in turn be detected in the blood serum of these "sacrificial lambs". They then injected the serum into the anemic sheep and looked for an increase in the number of blood cells in their blood streams. Unfortunately there were no visible results, even though the experiments had already been going on for more than a decade. It was obvious that the team had reached a dead end. When they were already on the edge of despair another research group fortunately published its discovery, revealing that excess EPO was not to be found in blood, but in urine. After fifteen years' hard work it must be quite depressing to find out that you have been looking in the wrong place all along. Nonetheless, they at least had a new goal.

A stroke of good fortune from the Far East

The lucky side of this misfortune was that a Japanese researcher, Takaji Mijake, offered Goldwasser help in collecting urine samples from anemic patients who supposedly produced excess EPO naturally, as a result of their illness. In just a few years, he collected and dehydrated 2550 litres of urine. In 1975, when the two scientists first met in the lobby of a Chicago hotel, Mijake bowed solemnly to Goldwasser and handed him a large package bound up in Japanese silk. Inside the neatly wrapped container was a priceless treasure, at least for the two scientists: a great amount of dehydrated, powdered urine. After a complex process of purification, Goldwasser and his colleagues managed to isolate 8 milligrams of pure human EPO from the collected urine. In August 1977 they revealed their discovery in scientific literature. Goldwasser and his team were overwhelmed with joy.

The goose that laid golden eggs

The story does not end here, though. In order to use EPO as a medicine it was necessary to find a way to produce it in large amounts outside the human body. Goldwasser was confident in the potential of his discovery, but at first no one took him seriously. He went from company to company, from one investor to the other, but did not succeed in getting anyone's attention. Fortunately, his search coincided with the beginning of the biotechnological revolution. It was at that time that pharmacists started to produce synthetic insulin with the help of bacteria cultures to cure diabetes patients. The more far-sighted entrepreneurs gradually came to understand that investing in biotechnology would earn them very good money. Some of them are enjoying the huge profits of that foresight today.

After many difficulties, which spread over the years from the fields of science to the courts of law, EPO became one of the first biotechnological geese to steadily lay golden eggs. Unfortunately, heaps of money do not always lead to new summits of scientific discovery, but more often encourage to new ways of mounting the profits still higher. After the huge commercial success of EPO in the 1990s, Amgen was said to have transformed quickly from a powerful scientific group into an outstanding law firm that just happened to hide, somewhere deep in its cellars, a department for medical research.

Cell police

Once upon a time, long ago when our planet was still uninhabited by plants and animals, single-celled organisms developed an effective method to defend against attacks from troublesome viruses that even then caused epidemics. It is only a few years ago that scientists discovered that the same mechanism single-celled organisms had developed as a defense system against viruses several millions of years ago, is still active in most living beings inhabiting Earth today, including people. A deeper insight into the workings of this mechanism could enable us to treat severe diseases more successfully. That is why it came as no surprise when the 2006 Nobel Prize in Medicine was awarded to the scientists who had, several years earlier, discovered how this primeval but still very effective cell mechanism, built into most of our planet's organisms, worked.

Like a country with strict regulations

The living cell is a highly complex mechanism in which many precisely orchestrated chemical processes take place. It could be compared to a country in which everything happens strictly according to exact regulations. Carefully stored in the central palace of the country's capital is a massive collection of laws and regulations containing exact regulations for everything that may happen within its borders. There is only one version of the "code", so it can never leave the central palace. However, the inhabitants of the country need precise instructions in order to produce anything, so a large number of scribes is constantly at work in the palace, transcribing dictates from the code and sending them around the country. When one of these transcriptions is delivered, the citizen who receives it immediately goes to work, following his orders to the letter. If the citizen in question is a cook, he makes the dish just as it is described in the recipe.

This metaphor roughly describes the way a living cell functions. The unique legal code containing all the information necessary for the functioning of the country is of course a metaphor for the genetic code which is carefully stored in a molecule of DNA and never leaves a cell's nucleus. The copies of instructions for building individual products and carrying out certain tasks are molecules of messenger RNA (mRNA) that carry information from the nucleus to different parts of the cell, while the diligent scribes making copies of the code in the central palace are enzymes called RNA polymerases (the 2006 Nobel Prize in Medicine was awarded for the research of this enzyme).

The copies containing the recipes, or molecules of mRNA, can easily leave the cell nucleus and move freely throughout the cell. As mentioned before, when the recipe reaches a cook, he reads it and prepares the meal exactly as it is written down. In actual cells, the workers diligently carrying out their tasks just as the Code of the land requires are called ribosomes. Their job is to use the information contained in the messenger RNA to build a certain protein which, in turn, performs specific tasks inside the cell. This is how most of the cells of all living beings on Earth work.

When a cell is attacked by terrorists

However, even a country in which everything happens strictly according to regulations has its faults and is open to attack. Evil terrorists can smuggle in their own malign instructions to take the place of authentic dispatches from` the official code. When a technician out in the country happens to receive a set of such "terrorist" instructions, he has no way of telling that they are not bona fide. . Hence he has no difficulties following terrorist dictations. Unconsciously, he works for an enemy of the state and can, without even knowing it, build a deadly murder weapon. These terrorists that infiltrate cells and abuse their mechanisms are known as viruses. They are extremely cunning and well-adjusted to living their piratical life.

Anti-terrorist squads

In response, cells quickly developed a method to fight such malevolent attacks. When terrorists enter a cell they first try to multiply or, in the terms of our metaphor, to issue as many instructions for creating a murder weapon as possible, and spread them throughout the cell. However, as they multiply, these packages of molecular data spend a few moments in the form of a double RNA helix. This structure is dissolved when the duplicate breaks away from the original "text", to be borne away and spawn copies of its own. And it is at this crucial moment, when the original and its copy are still glued together in their helix, that a cell can detect the fraud. It realizes that the infiltrated instructions are potentially dangerous and takes action: it sends special cellular police units to remove the bogus data. The anti-terrorist cell police does not only destroy the original instructions with the copy attached, but also tracks down and erases all other copies which might be present in the cell.

The technical term for such anti-terrorist units is RNA interference or RNAi. For their fundamental contribution to understanding this mechanism of cell resistance against viruses, the Americans Andrew Fire and Craig C. Mello were awarded the 2006 Nobel Prize in Medicine. RNAi is also an effective mechanism for controlling the way cells follow or ignore the dictates of individual genes.

The petunia mystery

As with so many important scientific discoveries, the first encounter with the RNAi mechanism was completely accidental. In 1986 a small California biotech company wanted to create a special flower. They tried to create a petunia that would have extremely vibrant violet petals. They knew which gene produced a violet color, so they tried to enhance its expression by injecting more instructions for creating violet color into the plant's genetic code.

The results of the experiment were surprising. Instead of becoming even more intensely violet, the flowers of this genetically modified petunia were now completely white. Naturally, the researchers thought they must have made some mistake, but after carefully reviewing the entire process found out that there was no error. The answer to the question of why the petunia flowers turned completely white and not more violet than usual became a mystery that took many brilliant minds more than a decade to solve.

Unlimited anti-terrorist awareness

Today, we know that when the new genes were being injected into the plant, the RNAi mechanism was turned on, interpreting the artificially added genes as an enemy virus attack and so destroying all molecular instructions similar to those that the scientists had tried to infiltrate into the cell. Because the newly added instructions were identical to those the cell had already produced itself, the cell's anti-terrorist squad started destroying both the artificially injected recipes and the ones that had been naturally transcribed by the "scribes" at work within the cell nucleus. The cell police was successful in destroying all the recipes for producing the color violet, so the petals turned out white.

For now, the discovery of the RNAi method has proven to be useful mostly in scientific research of individual genes, but it will soon become important in the treatment or at least in the diagnosis of various severe diseases. The cell's anti-terrorist squads can be taught to destroy specific recipes within cells which contain the instructions for the production of individual proteins. A number of diseases occur exactly because cells produce an excess of a specific cell product. Different methods are already being tested to use the RNAi mechanism in treating the various forms of hepatitis, Huntington's disease and AIDS as well as some forms of cancer.

The Black Death pandemic

Between the years 1347 and 1351, in just 1000 days, a mysterious disease killed nearly half of Europe's population. It was the greatest natural disaster to have ever befallen humanity. In terms of the number of victims and general devastation only the Second World War comes close to being directly compared with the medieval pandemic of the Black Death. This mysterious disease brought to Europe by seafarers was the bubonic plague. It started in Italy and other Mediterranean coastal towns and quickly spread all over the continent. It was caused by *Yersinia pestis*, a bacterium that is believed to have originated in China or India, and was transmitted to humans by flees found in rats. At the time, however, this was not yet known and only increased mass fear at the mysterious agent of death.

The mass murderer of Asia

The exact whereabouts of the source of the Black Death pandemic remain uncertain. It is very likely that this infectious and highly contagious disease first appeared in Mongolia and spread rapidly throughout China. Italian chronicles of the time contain a story which states that the infection was brought to Europe on Genovese merchant ships coming from the city of Kaffa by the Black Sea where Genoa had established a trading settlement. Kaffa was a rapidly growing city that lived off trade between the Far East and Europe. From there, Mediterranean ships transported exotic merchandise, brought from China by caravans, to European buyers.

According to the writings of a Genovese chronicler the trade settlement was attacked and besieged by the Tatar in 1346. When they came from Asia, they also brought the disease which swiftly spread through the ranks of soldiers. For this reason their commander gave the order to catapult infected bodies over the city walls. If this story is true, it could well have been the first occurrence of biological warfare in history, which at the same time also bears witness to the unpredictability and dramatic consequences of using such a weapon.

Genovese merchants inhabiting the settlement boarded their ships and escaped, carrying the deadly disease with them. The ships first anchored in Messina on the island of Sicily, but they were forced to sail on towards Genoa. Regardless of whether the story of the disease-carrying ships is true or not, it is a historical fact that the infection reached the Italian Peninsula at the end of 1347. By January 1348 the infection had spread to the South of France, and appeared in Paris that summer. It made its way to England early in 1349. Scandinavia was not affected by the infection until 1350 and Russia only followed a year later.

On all sides is sorrow; everywhere is fear

Between the years 1000 and 1250 Europe experienced significant economic and demographic growth. Around the year 1000 only about 40 million people inhabited the area of today's Europe. By the middle of the 14th century when the plague broke out, the population had nearly doubled, amounting to 75 million. Bearing in mind the medieval technology used for food production, the number of inhabitants at the time was already bordering on overpopulation.

According to various estimates thirty to fifty percent of all of Europe's inhabitants are believed to have died in only a couple of years as a result of the Black Death. Two thirds of the population of Florence perished in only a few months. The poet Francesco Petrarca (Petrarch) survived the pandemic, but lost his beloved Laura to it. In a letter to his brother, he describes the horror he experienced when he realized that he and his dog were the only survivors of the thirty-five people in the monastery where he lived.

> Alas! my beloved brother, what shall I say? How shall I begin? Whither shall I turn? On all sides is sorrow; everywhere is fear ... without the lightnings of heaven or the fires of earth, without wars or other visible slaughter ... well-nigh the whole globe, has remained without inhabitants ... houses were left vacant, cities deserted, the country neglected, the fields too small for the dead ...

In a highly scientific research paper entitled the *Compendium de epidemia per Collegium Facultatis Medicorum Parisius*, the professors of the Faculty of Medicine at the University of Paris attributed the cause of the plague to an event that took place on March 20th 1345. This was the day of the triple conjunction of Saturn, Jupiter and Mars in the 40th degree of Aquarius, which was supposedly a bad omen. The conjunction of the planets was believed to have caused a deadly decline in air quality which could have disturbed the balance between various bodily fluids. The main advice these eminent doctors gave the general public was to stay away from the infected air. Despite the now strange-sounding astrological theory on which this recommendation was based, the scholars' advice was actually quite appropriate.

The plague could end a life in less than three days. According to some witness accounts people would go to bed completely healthy, but by morning their hours would already be numbered. Doctors sometimes caught the infection from their patients and soon succumbed to the disease themselves. Giovanni Villani, a Florentine historian who undertook the task of chronicling the spread of the disease is stated to have died while writing the sentence: "And this plague lasted till –" Of course, he presumably did not die at his writing desk, but it must be true that he intended to finish the sentence at the end of his account, once the plague passed. Unfortunately, he never lived to see this moment.

Anti-Semitism and unusual religious sects

As the true cause of this sudden disease remained unknown, people quickly came up with different explanations for the catastrophe. Many were convinced that the plague was God's punishment for the sins of individuals as well as those of the Church which was preoccupied with accumulating wealth, giving feasts and other worldly matters.

The rapid spread of the plague across Europe also gave birth to unusual religious movements. In central Europe, the flagellation sect gained popularity. Even before the pandemic some monks practiced flagellation or other forms of self-mortification in their monastery cells in order to purify themselves of sins. After the arrival of the Black Death, the ceremony of flagellation had already turned into a public spectacle. Groups of fifty to five hundred men wrapped in white bloodstained cloths toured the villages and towns of Central Europe, whipping themselves while chanting at the top of their voices.

Yet he flagellation sect's attitude towards sinful Christian was positively lenient beside its view of the Jewish community. It was extremely anti-Semitic. During the peak of their popularity, flagellation groups in the area of today's Germany demolished a great number of the towns' Jewish areas. Jews were accused of deliberately poisoning water wells. It is true that mortality caused by the plague was lower in Jewish communities than in other parts of the population, but the reasons for this were mainly a higher standard of hygiene, on the whole, among Jews and the relative isolation of the Jewish quarters of towns and cities. All across Europe several hundred cases of Jew killings were recorded during the population's migration to northern parts of Europe, notably Poland and Russia.

Could the medieval plague have caused a minor Ice Age?

The Black Death had a severe effect on European society. It led to significant social and cultural changes. The medieval social structure had to adapt to new circumstances. There was suddenly a shortage of workers who were once redundant and badly paid. Compensation for the same work was now several times higher.

The fact that social circumstances can undergo such significant changes after a catastrophe of these proportions is not at all surprising. What about the impact on the natural environment? The fact that human activity, especially with greenhouse gas emissions, actively affects Earth's atmosphere is no secret. However, Thomas van Hoof from Utrecht University in the Netherlands formed an interesting hypothesis according to which humanity is supposed to have interfered greatly with the balance of the Earth's atmosphere in the Middle Ages. The theory is quite controversial, making it even more interesting. Van Hoof succeeded in estimating the concentration of carbon dioxide in the atmosphere over the past few centuries. He noticed a surge in atmospheric carbon dioxide between 1200 and 1300, while around 1350 the process suddenly reversed. The concentration of carbon dioxide in the atmosphere started to decrease.

Van Hoof advanced the hypothesis that, because of the disappearance of nearly half of the population of medieval Europe and Asia, a large number of fields were no longer in use and became overgrown with forests. This would have caused a significant amount of carbon dioxide to be absorbed by the vegetation that appeared in areas previously kept as fields and pastures. Carbon dioxide in the atmosphere forms a sort of protective or insulative cover for the Earth, preventing it from cooling down. Higher concentration of carbon dioxide in the atmosphere means better insulation or protection for our planet. If the amount of carbon dioxide in the atmosphere is decreased this means the Earth has shed one of its protective layers and has begun emitting more warmth into the outer space. This causes the temperature on Earth's surface to drop. According to this unusual hypothesis the medieval plague not only led to substantial social changes, but also resulted in a minor Ice Age (a drop in average temperature) which lasted from the end of the Middle Ages to the middle of the nineteenth century.

When a new, unknown disease breaks out

On February 28, 2003, the local office of the World Health Organization in Hanoi, Vietnam, received a call from a small private hospital with a capacity of no more than 60 beds. Two days before, its staff admitted a patient showing symptoms of atypical flu. To rule out a potential case of "bird flu" they requested the help of WHO's experts to try and determine what the disease really was.

Their call was answered by Carlo Urbani, an Italian-born contagious disease specialist and Doctors Without Borders veteran who, as the president of the Italian section of this important medical society, was a member of the delegation which received the Nobel Peace Prize in 1999.

"We do not know what kind of disease it is, but it is not flu."

When Urbani, as the official representative of WHO, examined the patient named Johnny Chen, an American-Chinese businessman, it quickly became clear that the situation was very serious. He suspected the unfortunate businessman to have contracted a completely unknown disease that doctors had never heard of before and were therefore unable to tell how dangerous and contagious it was. Urbani and the hospital's medical staff spent the following days collecting different samples and any other pieces of information they could gather from the patient, reviewed them and then sent them on to WHO, requesting immediate advice.

Urbani also made sure that a special secured, quarantined department was set up within the hospital. This soon turned out to be an important decision as one of the first conclusions the doctors were able to draw was that they were dealing with an extremely contagious disease. It was later found out that of the illness' first 60 victims half were members of the medical staff. When doctors first presented symptoms of the disease they faced the difficult obligation of isolating themselves as well to avoid infecting their families and so spreading the disease across the city. So they stayed in the hospital for the entire duration of the research. In one of his reports to a colleague, Urbani said: "I am in a hospital full of crying nurses. People are running around, yelling, and are completely frightened. We do not know what kind of disease it is, but it is not flu."

None of the precautions taken by the hospital's medical staff turned out to be excessive. It soon became obvious that this was a case of a new viral disease which was not only highly contagious, but also very dangerous. They were unaware of this during the first weeks, but statistics later showed that one patient out of ten died. On March 9 WHO had already gathered enough information to meet with the highest representatives of the Vietnamese authorities and warn them about the gravity of the situation. Professional help from the international community had by this time begun reaching the hospital. Experts from abroad brought with them all the equipment developed and used in research of the most deadly and contagious diseases such as the Ebola virus. The private hospital was shut down and its patients were transferred to a special department of the Bach Mai public hospital where local doctors teamed up with members of Doctors Without Borders who were already used to dealing with similar situations.

When all the key precautions for fighting epidemics had been taken, the death rate stabilized. The outbreak of the disease in Vietnam became a good example of how to act when such a disease is suspected to have occurred. If Urbani had not been as convincing in warning the authorities to react as quickly and transparently as they did, we could have witnessed a catastrophe of epic proportions.

On September 11, soon after the situation in Hanoi was at least partially under control, Urbani flew to Bangkok to take part in a science conference. While still on the plane, he felt ill and already started showing typical symptoms of the newly discovered disease. A colleague was waiting for him at the airport, but he did not let him come near as it was obvious that the virus had attacked him as well. For over an hour, Urbani and his colleague sat quietly each on his side of the waiting room, waiting for the ambulance to bring all the equipment necessary for doctors to protect themselves from dangerous infections.

Urbani was taken straight from the airport to quarantine in a local hospital, where he spent the next 18 days fighting for his life. He developed the respiratory symptoms typical of all the virus's other victims. He could only talk to his wife and three children over the phone because the infection was so dangerous that no one without adequate protection was allowed to come in direct contact with him. Despite his extensive knowledge of infectious diseases and the help of medical colleagues who flew in from Germany and Australia, bringing with them several new antiviral drugs, he finally lost his battle with the new and deadly disease. He died on March 28, 2003, a month after his expert advice was requested by the hospital in Hanoi. He left his diseased lung tissue to science.

The great Hong Kong virus carrier

Johnny Chen however, the businessman who brought the disease to Vietnam, was not the first to have been affected by the virus. WHO soon received information from other parts of the world about this new, highly infectious form of flu that quickly became known as SARS. In no more than a few weeks the disease had spread to three continents and it seemed that humankind was threatened by a pandemic of unforeseeable proportions. That is why on March 15 the WHO Director-General issued an alarming warning in which he urged that strictest precautions be taken in order to stop the pandemic.

Researchers later found out that the first case of SARS probably occurred in November 2002 in a Chinese boy from Foshan. On the November 16, 2002 he was admitted into the local public hospital with an atypical respiratory disease. How he had contracted the disease remains unknown, but we do know that he recovered after infecting several others who quickly spread the disease first across China and Hong Kong, then across the globe.

In China, the first "great disease carrier", as epidemiologists call individuals who, due to their lifestyle or workplace, are likely to infect a large number of people, was a fish merchant called Zhau Zuofeng who caught the virus in January 2003 in Guangzhou. He not only transmitted the disease to the staff of three hospitals (at the end of the epidemic it was found out that 20% of all the infected were medical staff), but also to a nephrology professor who had just left for Hong Kong. There he stayed in room 911 on the ninth floor of the Metropole Hotel. After only ten days, the elderly professor succumbed to the disease after infecting several other guests of the hotel who were also staying on the ninth floor. In turn, they carried the disease to Toronto, Singapore and Vietnam. It was during his business trip from Shanghai to Hanoi that Johnny Chen, the man who carried the disease to Vietnam, made a stop in Hong Kong and spent the night on the ninth floor of the Metropole Hotel.

On March 15 when WHO issued a warning urging all travelers to be extremely cautious, no one knew much about the disease, except that it was a highly contagious form of atypical pneumonia. At first, experts suspected it was a modified flu virus, the symptoms of the disease indeed implied, but tests did not confirm this hypothesis. Because the lungs of the deceased were extremely damaged, suspicions arose that it might even be a case of lung plague, but as antibiotic treatment had no effect, this hypothesis was rejected as well. Soon, scientists agreed that SARS was the first serious new illness discovered in the 21st century.

A record-breaking discovery of the cause

On March 17 WHO assembled a team of the best microbiologists, virologists, epidemiologists, and clinicians in the world to battle against the new disease. With conferences being held daily and all the information transmitted via the Internet, it was established at the beginning of April that the disease was caused by a new virus of the coronavirus family which had never before been noticed in humans or in animals. Coronaviruses are relatively harmless and cause common colds. SARS, it seemed, was no typical coronavirus. By April 12, the scientists were already familiar with the entire genome of the virus and on the May 1 an article was published giving a detailed description of the virus.

In the middle of May, when the epidemic reached its peak and each day was filled with reports of several hundred new cases, scientists were still without any medication or vaccine to fight it, so the authorities had to resort to usual measures that have been practiced by humankind for thousands of years. All those contaminated were immediately isolated to prevent the disease from spreading further. In Singapore, webcams were installed in the homes of potential carriers of the virus, to ensure they kept the terms of their quarantine. The penalties for leaving quarantined homes were very severe. In Hong Kong, the apartment building in which the highest number of infected people lived was evacuated and its residents were moved to a special camp for ten days.

In China, it took longer for the authorities to realize the gravity of the situation. In the end, this was the reason why a quarter of all reported cases of SARS were from Beijing. Of course, it could have been even worse if, at a certain point in time, the Chinese authorities had not reacted so firmly. They shut down all the schools, theatres and cinemas, and prohibited all public events.

At the end of April they decided to build a new hospital in the suburbs of Beijing which would specialize in treating SARS. Seven thousand construction workers set a record by constructing the hospital in no more than eight days for 170 million dollars.

Then, as suddenly and mysteriously as the disease appeared, it began to vanish in the summer of 2003. Considering the fact that 774 people died of SARS, it was fortunate that only 8,096 people were infected.

A deadly virus from the heart of Africa

At the end of August 1976, Mabalo Lokela, a forty-four-year-old teacher, returned from his travels, with a high fever. He had been travelling in the north of Congo near the border with the Central African Republic. The medics who examined him in the mission hospital in his home town of Yambuku suspected that he had contracted malaria, so he was given a shot of chloroquinine. His temperature came down, so he was released from the hospital and left to home care, but after only a few days the fever came back.

In the beginning of September his family brought him back to the hospital, but by then he was already in very poor condition. He vomited uncontrollably, had acute diarrhea, headaches and trouble breathing. In the days that followed his nose, eyes and gums began bleeding. Unfortunately, there were no doctors in the mission hospital, only nurses, so although they did everything in their power to help him they were unable to diagnose or treat the disease effectively. On September 8, approximately two weeks after presenting the first symptoms of the illness, Mabalo died.

"We're all going to bleed to death!"

In keeping with tradition, Mabalo's wife Mbunzu, who took care of him all along, washed her husband's body with her mother, her husband's mother, her sister and other close female relatives to prepare him for his funeral. Soon after the funeral, they along with other relatives and friends who were present at the ceremony all developed the same symptoms as those the deceased teacher had died from only a couple of days earlier. Some nurses from the hospital where Mabalo was being treated also fell ill. People started to panic; it seemed that they were all going to bleed to death.

On September 15, Dr. Ngoy Mushola, a doctor from the nearby town of Bumba, came to Yambuku at the request of the director of the mission hospital. He realized quickly that they were dealing with a case of a highly contagious disease, one that was already spreading rapidly through the town and the neighboring villages. He wrote down a detailed description of the symptoms and tried imposing at least a minimal quarantine upon afflicted people. This was necessary because the locals were used to burying their dead right next to their houses or even inside their dwellings, and this only accelerated the rate of infection. On 17 September Dr. Mushola sent his first official report of the outbreak of this contagious disease to those responsible in Kinshasa and requested immediate assistance.

On 23 September two experts on microbiology and epidemiology, otherwise professors at the National University of Zaire, came to Yambuku to assess the situation. After they arrived, they diagnosed the illness as an outbreak of atypical yellow fever and soon left the area of epidemic. They brought a very ill nurse and two other members of the staff from the mission back with them to Kinshasa. The samples of the diseased nurse's blood were then sent to the Pasteur Institute in Paris through the French Embassy in the hope of a precise diagnosis. Analysis of these samples at the Institute revealed that the disease was caused by a completely new virus. The disease was named Ebola after a stream in the vicinity of Yambuku from which it was thought to originate.

Shortly afterwards, at the beginning of October, the local authorities brought in the army isolate Yambuku and the surrounding area with road blocks. Movement was also restricted in the larger nearby town of Bumbu, two field hospitals were set up as quarantines, and travel to other parts of the country was prohibited. Despite these efforts, the disease continued spreading to many nearby villages, infecting 318 people of which 280 (almost 90%) died.

At the same time, a slightly different strain of the virus spread from this area in the heart of Africa to the south of Sudan. The Sudanese manifestation of Ebola was slightly less deadly, with a mortality rate of just over 50%. But still, out of 284 people who caught the disease 151 died.

The monkey connection

Virologists studying this strange and frightening new killer illness in northern Congo and southern Sudan presently discovered that the virus was similar to one that had appeared in Europe barely a decade earlier. In 1967 in Marburg, Germany, the local clinic for infectious diseases had admitted a number of severely ill patients running very high fevers, suffering from great pain and eventually bleeding from several parts of the body. It turned out that all of them worked at the same pharmaceutical company and had been infected by a virus they had contracted from monkeys imported from Uganda. Half of the monkeys, which were needed to prepare the cell cultures for use in producing vaccines, had already died on the way from Africa.

Apart from Marburg, some cases of infection with the same virus were also discovered in Frankfurt and Beograd. Twenty-five people were infected by direct contact with the monkeys and seven did not survive. Besides these primary infections, six more individuals fell ill, but there were no more deaths.

When it comes to outbreaks of this deadly viral disease, the one fortunate thing is that the Ebola virus kills too rapidly to spread effectively among people. The infected individual quickly becomes too weak and ill to walk around spreading the infection, so Ebola viruses are thankfully limited in scope. From 1976 less than 2000 people contracted Ebola, but the mortality rate of each outbreak was as high as 50-90%. One of the highest death rates was recorded during the 1976 outbreak in Congo.

Could bats be the carriers?

Although we are much more familiar today with the ways such fatal diseases function and spread than we were decades ago, in African countries outbreaks of Ebola and similar viruses still occur. In 2007, around a hundred people were afflicted by Ebola, a quarter of whom died. In the middle of July 2008, a Dutch woman who had returned from her travels in this African country also died from an Ebola-like virus. She was thought to have been infected by a bat during a visit to a cave.

On the HealthMap website (www.healthmap.org) a map of the world enables us to keep track visually of the latest reported outbreaks of disease. The website automatically tracks professional health information sources and warnings as well as newspapers and other news sources, classifies outbreaks according to the level of danger and locates them on the map. At the beginning of this summer, a red flag marked the Dutch city of Leiden where local doctors wearing protective gear in a specially secured part of the main hospital were treating the infected tourist mentioned above.

Humans are not the only species in need of a vaccine against Ebola, since it not only attacks people, but can also affect monkeys and forest antelopes. The disease poses a significant threat, for example, to the gorilla population. There is thus an element of environmental conservation as well as public health in the fight against Ebola.

One of the more important questions concerning Ebola and similar diseases is what species is their original natural host and how the viruses are then transmitted to other species. For some time now it has been suspected that the natural carriers of this disease are bats, but the hypothesis has yet to be confirmed. However, it has been established that bats are not affected by the disease even if they carry the virus. One of the first cases of infection in Sudan in 1976 was an employee of a cotton factory which contained a large number of bat-nests. It has also been suggested that forest-dwelling animals might contract the virus by collecting and eating fruits that have already been chewed on and cast off by bats. The teacher Mabalo Lokela, who was officially the first to have died from Ebola was thought to have eaten antelope meat during his travels in the northern part of Congo.

For exceptionally dangerous diseases such as Ebola or anthrax, the American Food and Drug Administration (FDA) even changed its rule of only approving a vaccine once it has been successfully tested on people. For these especially lethal diseases vaccine production may go ahead if scientists prove that an experimental drug successfully immunizes two species of animals. Naturally, it would be difficult to find someone who would be brave enough to take an injection of such a prototype vaccine and then be voluntarily infected just to prove that the vaccine really works.

Malaria

The World Health Organization estimates that several hundred million people are infected with malaria each year, of which more than a million die. Science has been waging war against this disease for more than a hundred years. A few decades ago it seemed that victory lay ahead, but malaria adapted and struck again.

The deadly "mosquito conspiracy"

Malaria can kill a man in a matter of days. The disease is caused by a small, single-celled parasite, Plasmodium, which inhabits the red corpuscles of the infected person and starts to divide, doing so until the cells bursts. Destroyed blood cells can clog blood vessels causing damage to organs which may lead to the death of the affected person. It is most likely that, throughout the history of mankind, this tiny parasite has killed more people than any other disease.

For a long time people believed that malaria was caused by infected air. Hence its name: male aria, in Italian, means "bad air". The key discovery in understanding how disease really works was made in 1897 by a British military doctor stationed in India. Dr Ronald Ross was testing the hypothesis that the true carrier of malaria was not bad air but the mosquito. The age's leading authorities on the disease dismissed the mosquito theory as absurd, but Ross decided anyway to put it to the test. He conducted an experiment that proved to be one of the most important in the history of medicine. He bred some mosquitoes and fed them blood taken from infected individuals. He wanted to find out if a malarial parasite can even survive inside a mosquito host. He tested his theory on several species of mosquitoes, but the parasite always perished. That was until he discovered that the parasite does in fact thrive inside the digestive system of one particular species of mosquito, of the Anopheles genus. Not only does the parasite survive inside this insect, it reproduces and waits to be transmitted to a human victim.

Ross's findings, for which he was awarded the Nobel Prize in 1902, gave the human race its first real chance of fighting malaria. The idea was simple: as malaria was transmitted by mosquitoes, it was necessary to change the natural environment in a way that would make it as inhospitable for mosquitoes as possible. Extensive projects were started, draining swamps, stagnant waters and slow rivers where mosquitoes were most abundant. For the first time in history, the number of malaria infections started to dwindle. However, the method of draining was not as effective in certain tropical areas of the world, so the search for a more powerful weapon began.

Chemical weapons enter the battle

When fighting in the tropics, armies were horrified to find out that in these areas more soldiers were killed by malaria than by bullets, making it imperative to find an effective method of controlling the disease. During the Second World War, the Americans developed an insecticide that effectively destroyed the mosquitoes' nervous system even when applied in small doses. After being dispersed over an area, it remained effective for several years. In addition to that, it was cheap. The insecticide was called *dichloro-diphenyl-trichloroethane* or DDT. After the war, the World Health Organization decided to use DDT to eradicate Malaria completely over the next few decades.

At first, the plan worked like a charm. In India alone, the number of malaria infections decreased from eight million cases before the eradication began to a "mere" fifty thousand. It seemed that the battle with malaria had already been won when mosquitoes struck again. Their weapon against DDT was evolution. Every day, millions of mosquitoes come to life all over the world, and occasionally, some appear that have different traits than the majority. It only took one mosquito to mutate in such a way that it became resistant to DDT. The insecticide had no effect on this particular mosquito, so it was able to reproduce without any difficulty and soon a population of DDT-resistant "supermosquitoes" was rampant across the old malarial regions.

Chemists put their heads together and started developing new insecticides, but mosquitoes eventually became resistant to those as well. The war on mosquitoes was a war on evolution itself. New insecticides had to be developed faster than mosquitoes could develop new defense systems through random mutations which kept the species from becoming extinct. As a rule, however, every new insecticide is more expensive than the previous one. In 1969, the World Health Organization gave up its struggle against mosquitoes and called off its plan to eradicate the chief carrier of malaria. Once more, the mosquitoes multiplied and the number of deaths caused by malaria increased again. The number of infections in India leaped from a couple of thousand to a couple of million.

Because of mutations, the malarial parasite also became resistant to once relatively effective medicines, such as quinine, brought from Peru by the Jesuits in 1640. Some twenty years ago it therefore seemed that humanity had lost its battle with one of the deadliest diseases on the planet.

An ancient Chinese remedy supplies a cure

In the 1970s nobody in the West knew that in fact the Chinese had developed a very effective remedy against malaria. In the spirit of the Cold War, however, this had been kept top secret. The cure was discovered when medical researchers – allegedly following Mao's advice – carefully studied more than 200 traditional Chinese treatments for malaria, one of which proved incredibly effective. Among the traditional recipes they tested, was a 2000-year-old set of instructions for preparing Quing Hau Su tea. The scientists were amazed by its curative properties. On analyzing the tea's ingredients they found that the active ingredient that possessed the tea's healing effects. Today, this substance is called artemisinin and is the most powerful antimalarial drug that has yet been discovered. Because of the Cold War and the concealment of the discovery from the world, it was thirty years before the medicine became internationally available on a large scale.

News of the miraculous antimalarial drug reached scientists from beyond the borders of China through an article published in a Chinese medical review, a source which at the time did not inspire much confidence. In addition, the Chinese authorities did not allow the ingredient to be analyzed in any laboratory outside China. It took several years for westerners to find the plant that produced artemisinin and test the medicine for themselves.

How to produce enough medicine for all the infected?

Because the drug turned out to be very effective, the demand for it today largely exceeds the available supply. When *Artemisia annua* is planted it is another eighteen months before artemisinin can be isolated from it. This makes it impossible by conventional methods to produce the drug in sufficient amounts to answer the needs of all those infected with malaria. For this reason scientists today are striving to develop a method of manufacturing the drug more quickly and efficiently.

By transferring some of the genes responsible for the production of artemisinin in the *Artemisia annua* plant into yeast plants, researchers have succeeded in making microorganisms produce artemisinic acid: a substance which, by means of a further chemical procedure, can be used to yield the actual drug, artemisinin. With the help of yeast plants and this procedure it will most probably soon be possible to manufacture much larger quantities of the cure and save countless human lives. In Africa alone, two children lose their lives to malaria every minute.

As with every new drug, however, caution is necessary. In January 2006, the World Health Organization appealed to the pharmaceutical industry to stop advertising and selling drugs based solely on artemisinin. Last year, researchers realized that the mutation of a single amino acid could be enough for resistance to reappear. One of the best strategies in preventing malaria from becoming resistant to the latest, most effective solution is by combining it with other antimalarials in treating patients. This reduces the possibility of a mutated form of the disease surviving and then spreading across the world, thereby preventing a disaster which has already occurred on several occasions.

The child prodigy who became the father of cybernetics

During the first years of the twentieth century rumors started going around the Harvard campus about the extraordinarily gifted child of the university's professor of Slavic languages and literature, Leo Wiener. The Wieners' son Norbert had learned the alphabet by the age of 18 months. At three years of age he read books from his father's library and at five he could recite Greek and Latin classics by heart. Soon he delved into chemistry, mathematics, physics and biology, and enrolled at Tufts College when he was only eleven, becoming the youngest college student in US history.

Harvard doctor at eighteen

On October 7th 1906, *The World Magazine* published an extensive article about the talented Norbert. The magazine interviewed the boy and also his parents, who stated reassuringly that they doing everything to give the boy as normal a childhood as possible. However, it was clear that without his parents' unyielding support the boy could never have been as successful.

Aside from literature and philosophy, the young scholar always had a lively interest in science. He was influenced in particular by his father's friend, the physiologist Walter Cannon, who is, among other things, famous for introducing the term homeostasis into biology. Norbert enjoyed exploring Cannon's Harvard laboratory and it is most likely that this acquaintance led to his decision to become a scientist rather than a philosopher.

After graduating, he first studied zoology at Harvard, then philosophy at Cornell and at the age of eighteen was awarded a doctorate for a dissertation on mathematical logic at Harvard University. In the following years he visited Europe and spent some time studying under Bertrand Russell and G.H. Hardy at Cambridge, and under David Hilbert at the University of Göttingen.

After the First World War he secured a position as a lecturer at MIT where he remained for the rest of his career.

According to a family legend, one of the ancestors of the Wiener family was the Jewish philosopher and physician Moses Maimonides who was himself, like Norbert, a child prodigy. Maimonides' descendants are believed to have lived in Poland from where the Wiener family originated, but the sources supporting this claim were lost in a fire, so the story is no more than a story, as Norbert confirmed in his writings.

Attempts to automate air defense

At the beginning of the Second World War, Wiener sent letters to the President's science advisor outlining in detail what was in effect a prototype of the modern computer. At that time, however, no attempt was made to make his idea a reality. When he learned about the horrific consequences of the German bombing of England he decided instead to put all efforts into finding a way of coordinating air defense as effectively as possible.

Military mathematicians were able to calculate very precisely how to direct artillery to hit a distant target. The problem with hitting an aircraft, however, was the speed at which the target moved. That is why Wiener tried to include all available data on the direction and speed of an aircraft in his calculations, thus improving the accuracy of anti-aircraft artillery. However, as these calculations were quite complex and took much time to make, he searched for a way of automating this process.

The main addition Wiener made to his model of "intelligent" air defense was radar, a new technology which had just been developed by the English, so that his calculating device could continuously receive accurate data on the position of an individual aircraft. His idea was reviewed by a special commission responsible for coordinating scientific work in wartime. The commission approved the proposal and directed that experimental construction should begin. The chief project engineer was a young Julian Bigelow, with whom Wiener immediately hit it off. First they approached the problem of shooting at airplanes theoretically, using complex differential equations, and then they visited several military bases where they could observe the flying capacities of military aircraft first-hand.

At first it seemed that predicting the positions of airplanes would be something quite impossible as pilots were able to maneuver them with great skill and avoid fire from the ground very successfully. However, their freedom in taking evasive action was limited by the great speed at which they flew. Airmen had to be careful not to accelerate too much when turning. With this in mind Wiener and Bigelow were able to determine quite accurately the range of positions through which an airplane could move.

The birth of cybernetics

The deeper Wiener delved into the problem of coordinating air defense, the more he believed he was on the path to discovering a universal logic that could in many cases also be found in nature. He started to develop ideas about feedback loops, the transfer of information and circular causation, and these ideas soon turned out to be important in dealing with the functioning of living beings as well as modern technology, especially in the fields of electronics and information technologies. Wiener would later name this field of science cybernetics.

Wiener's fundamental idea was that cybernetics did not only explain the fields of information technology, but the nature of man and human society as well. Information flowing between people, animals and machines is essentially the same. An efficient flow of information is of vital importance to the stable functioning of a system, as it enables it to maintain equilibrium or homeostasis by way of feedback loops.

The idea of cybernetics as a way of introducing new ways of thinking with technological promise for practical application has also drawn in many other eminent scientists. Soon after the war a narrow circle of these scientists started gathering, in special meetings which could last for several days, during which they exchanged thoughts on the vast array of different applications the basic principles of cybernetics might serve. Apart from Wiener and his MIT colleagues, the group included the all-round mathematician and physicist John von Neumann, some neurologists, biologists and the famous anthropologist Margaret Mead.

The scientists were united by the idea that the principles of cybernetics could enable them to learn how the brain and thought functioned. This would in turn give them the knowledge to develop an artificial brain, or a machine something like today's computer. At the first conference, Mead presented her research describing how some of the peoples of the South Pacific maintain social stability with rituals which she had interpreted as a form of social information feedback loops.

A critic of social development

Despite the fact that Wiener had always been actively involved in scientific projects for the American military during the Second World War, he was not invited to join the chosen few who were selected to develop the atom bomb. His scientific credentials were not at all in doubt, but military intelligence regarded him as a security liability. He was not very stable emotionally, and was a fervent supporter of public access to all knowledge, a conviction which caused him to part ways with some of his research colleagues after the war. Also problematic was his marriage to a German woman whose family back in Germany was loyal to the Nazi regime. Indeed a cousin of his wife's worked as an administrator in a concentration camp.

After the war, he quickly realized that society was not developing the way he had hoped. He was deeply upset by the fact that more and more knowledge was kept secret. During the cold war the government engaged many scientists to take part in its military projects. These scientists developed many different technologies, but a lot of the findings remained hidden from the public. Wiener had a vision of a stable world functioning according to the principles of cybernetics. This stability would be based on an efficient information flow, maintained by feedback loops, which would keep society in a state of homeostasis, much as a living being is kept alive by its inner regulatory mechanisms.

He voiced his concerns publicly, after which he was followed by secret services who suspected him of "subversive activity" and "sympathizing with communism". He only managed to avoid more severe sanctions because he had never worked on any of the military's most classified projects and could therefore not reveal any secrets about nuclear weapons or plans of similar strategic consequence. However, his professional career did suffer a blow as a result of his critical public statements. Life today would be almost unimaginable without a number of technologies made possible by Wiener's pioneering work on cybernetics. Even so, his name was soon largely unknown beyond the specialists who continue to benefit from his thought.

Statistics against poverty and disease

Not many people can read music notes well enough to hear the music in their minds just by looking at a score they have never heard played. Hans Rosling, a professor of medicine at the Karolinska Institute in Stockholm, claims that something similar applies to statistics. Pieces of raw data are like notes on paper, requiring a musical instrument and an interpreter so they can be understood by an average person. In fact, we can understand a great deal of only when it has been very clearly presented with analogies that we can visualize.

The visualization of human development

Just over a decade ago, when Rosling was preparing his first lectures on global public health for his Swedish medical students, he soon realized that they were having difficulties with visualizing the mass of statistical data which made up the basis for studying the level of economic and medical development in individual areas of the world. With the help of his son he therefore began work on a computer program which would present the enormous amount of information much more clearly. In this way he could breathe life into this data much as a musician who transforms the musical notation of a composition into sounds.

After a week or so, Rosling and son had developed the first, test version of the software they called Trendalyzer. They also had some help from his son's wife, who designed an attractive user interface. With this program, which they uploaded to the Internet and could be accessed for free, they could animate the chronological changes of the most important statistical data of all the countries in the world from the middle of the nineteenth century to this day. In 2005, after the program's great success, Rosling founded the non-profit Gapminder Foundation (www.gapminder.org), the mission of which was to create free-access applications for the visualization of human development. In March 2007, Trendalyzer's future development was taken over by Google, upon agreement that they would still offer the software for free to all internet users.

Rosling negotiated the software takeover with the famous Marissa Mayer who is, besides the company's co-founders, one of Google's most important people, having the last word in everything that concerns all new products the internet giant offers. Mayer immediately advised Rosling to add short and attractive video presentations of different applications his brilliant computer program could be used for. Rosling at first thought that this was overdoing it, because he was convinced that anyone getting to grips with the mass of information could bring it to life themselves with the help of the software, but he quickly admitted his mistake. It was evident that merely the data and the software for its visualization were not enough, just as music notes and a musical instrument do not make music by themselves. They need an interpreter and it turned out that, when it comes to statistics, Rosling is a natural. Rosling's short lectures, which he named GapCasts, are also available on YouTube and have proved very popular.

Today is different world

Rosling likes to joke that, at the moment, his software has two important target groups: children under the age of twelve and influential politicians. It is characteristic of both groups that their attention has to be captured in a matter of seconds or they lose all interest. Policy makers and administrators are almost always out of time, so they find this sort of visualization of statistical data very useful. After the Cold War and in the era of globalization, one has to be up-to-date with all kinds of information, because the standard means of classifying and analysing countries no longer apply.

The main message Rosling has been trying to convey to his students and in recent years to a wider audience is that today's world can no longer be described by slotting countries under out-of-date headings such as traditional "developed" and "developing" countries. By animating statistical data, Rosling has clearly illustrated that such divisions might have corresponded to the actual circumstances prevailing in the first decades after the Second World War, but that by now the world has changed completely. According to several criteria, many of the "undeveloped" countries for which experts held out few hopes only half a century ago, have in fact drawn close to "developed" countries.

Rosling makes an example of himself by saying that he only knows two sorts of wine: white and red. Similarly, most people have a simplified idea of countries, dividing them into two types: the developed and the undeveloped. However, just as wines come in a huge number of varieties, there are many different types of countries. This becomes all the more evident if one is willing to take a look at all the information available. Rosling prides himself on knowing no less than two hundred sorts of countries.

As an example of miraculous development, Rosling likes to present Bangladesh during the last three decades. Today, this Asian country with almost one thousand inhabitants per square kilometer is the most densely populated of all the larger countries of the world. India ranks second with a population density three times smaller. 150 million people live in Bangladesh, which is more than the population of Russia, while the area the country covers is no bigger than that of Florida.

In 1970, the women of this poor land east of India had an average of seven children and every fourth child died before he reaching the fifth year of life. At the time, the opinion of numerous experts was that large families are so deeply rooted in the culture and religion of this country that its huge ongoing growth in population was unlikely to change. However, after declaring its independence from Pakistan in 1971, the country's general indicators of the development in health began to improve drastically. In 2003, an average Bangladeshi woman had only three children in her lifetime and only one in fifteen children died before their fifth birthday. But the main point is that this change did not occur because of migration from villages to cities, but because the quality of life in the countryside improved.

The seemingly impossible is possible

Rosling also poses himself a seemingly unusual question: is the world today on average better than it was three decades ago? He finds the answer with the aid of the term "extreme poverty". In 1970, 38 % of all people (1.4 billion), which is more than a third of all humanity, lived under the limit defining extreme poverty, surviving on less than a dollar per day. Extreme poverty means waking up hungry in the morning, going to bed hungry in the evening and spending all the time in between spending energy to get food. In 1990, 26% of all people lived under the limit of extreme poverty and in 2000, 19%. According to projections, supposedly only 10% (0.7 billion) of mankind will live in extreme poverty by 2015. However, Rosling warns us that this data is relatively unreliable and therefore is not necessarily a completely accurate reflection of the present and past world.

Despite the unreliable data, most researchers still agree that during the last three decades the percentage of mankind living in extreme poverty has diminished by half. If, in 1970, two out of five inhabitants of the planet were hungry all day, in 2005, only one in five were extremely poor. That is how the idea that humankind can rescue its entire race from extreme poverty has now become a generally accepted opinion among the experts.

Rosling has already accepted numerous awards and was declared one of the top scientists of the year 2007 by Discover magazine. Rosling is not merely a theoretician playing around with numbers, but has spent many years as a doctor in remote locations of the world. At the beginning of his career he described a new disease in Mozambique and also discovered a successful treatment to fight it.

Apart from science, Rosling has a rather unusual hobby. He trains in circus arts. At the end of a now already legendary lecture at a conference in the US, and in support of his claim that even apparently completely impossible goals like ending poverty all over the world are actually not as unattainable as they might seem, he did something very unusual, especially for a scientist. He asked for a moment of silence, took off his shirt and, in the manner of an Indian sword-swallower, slid a meter-long Swedish army bayonet down his throat. He accompanied his performance with the following thought: "Sword swallowing is from ancient India. It's a cultural expression that for thousands of years has inspired human beings to think beyond the obvious."

The Pasteurization of heretical ideas

The French chemist and microbiologist Louis Pasteur was well known for presenting his research results to the public in the most extravagant fashion he could think of. He liked to stage great spectacles which would undoubtedly be broadcast live on television today, and which were covered in detail by mid-19th century newspapers. On June 2, 1881, Pasteur became a media star on making public the results of a bold experiment he conducted at a farm in the French countryside. When he had announced a few months earlier that he had discovered a vaccine against anthrax or splenic fever, using weakened *Bacillus anthracis* bacteria, the veterinarian Hippolyte Rossignol challenged him to conduct a controlled public demonstration of this controversial treatment. Rossignol opposed Pasteur's theory that anthrax was caused by microorganisms, and hoped that the experiment would discredit him.

Spectacle at Pouilly-le-Fort

Accepting such a challenge was of course a risky endeavor for Pasteur because it could easily hit an unpredictable complication and result in public humiliation. However, his passion for spectacles and public attention prevailed. The experiment was carried out under the supervision of the Agricultural Society of Melun, presided over by the Baron de la Rochette. On April 28 Pasteur signed an agreement with the society containing clear specifications as to how the experiment was to be conducted. He was given sixty animals of which one half would be vaccinated. The animals would then be kept under observation to see how they would react to the anthrax infection.

The experiment was carried out according to a strict and predetermined protocol at the Rossignol farm in the village of Pouilly-le-Fort near Melun. On May 5, twenty-four healthy sheep, one goat and six cows received their first shot of the anthrax vaccine. On May 17, the vaccination was repeated with the second dose of the vaccine. Finally, on May 31, all of the test animals were infected with a highly concentrated dose of bacteria which should normally have resulted in a very rapid development of the disease. Pasteur confidently predicted that only the vaccinated animals would survive the infection.

On June 2, 48 hours after the animals were infected, a great number of journalists, veterinarians and other personalities travelled to the rural farm to hear the results of the experiment. A remarkable spectacle was awaiting them: the twenty-four sheep, the goat and six cows that had been vaccinated seemed to be in good health, while twenty-one untreated sheep along with the other goat had already died. Two of the unvaccinated sheep died before the journalists' eyes and the last animal did not live to see the end of the day. The infected cows did not die, but displayed symptoms of the illness and suffered from an elevated temperature.

News reports sang praises to Pasteur, but no one mentioned that he was only one of many scientists who had participated in developing the vaccine. The method for treating anthrax bacteria for use in the vaccine, weakening the germs to stimulate immunity without bringing on a full attack of the disease was in fact the work of a number of experts. In his famed Pouilly-le-Fort experiment, Pasteur did not even use the vaccine prepared according to his own method, but instead used one prepared by a rival research group which happened to be more effective at the time. Nevertheless, it was Pasteur who saw how to piece such individual discoveries together; and he too who had the nerve to present the result in a public spectacle.

In a letter Pasteur's wife wrote to her daughter after his great success, she mentioned that apart from the test animals there was someone else who had died that day: "On the day when our twenty-five sheep perished, we received the news that Monsieur Littré had passed away as well." The prestigious Académie française has only forty members and it is not until one of them dies that another is chosen. Pasteur had justified hopes that, because of his resounding achievements, this time the academicians would elect him. He did not have to wait long before he was informed in December 1881 that he had been accepted to join the ranks of the "immortal".

What only living beings can do?

Pasteur had been taking on an active role in similar public issues since the beginning of his career. As a young researcher he studied the interaction of polarized light and crystals and discovered that tartaric acid crystals, also found in wine, were optically active. This means that when these crystals are exposed to light, they react in a certain way. He then made the bold hypothesis that only living beings can produce optically active substances. In her letters to her family his wife Marie proudly declared that Louis was conducting experiments that could make him the next Galileo or Newton.

Throughout his career, Pasteur also worked closely with industry to help solve problems with chemical and biological processes that caused difficulties to factories. As a professor at the University of Lille he tackled practical issues concerning the local food and brewing industries. He proved that fermentation or brewing was as much a biological as a chemical process. The brewing process involves the active participation of microorganisms, which have a vital role. He also invented a method of rapidly heating liquids to 60° Celsius, a temperature at which all microorganisms die, and are thus prevented from multiplying later and spoiling food or beverages. The process of Pasteurization, which was patented by Pasteur in 1866, is still used in the food industry today.

His capacity for public performance encouraged him to get involved in a famous debate with Professor Félix Pouchet over the possibility of the spontaneous generation of life. The subject was very close to the problems he had been dealing with in his research into the fermentation process, so he believed it provided the right platform for a public appearance.

At the time, the problem of a spontaneous generation of life was by no means merely a technical question, but was closely linked to questions regarding religion and morals. The possibility of a spontaneous generation of living beings from inanimate matter was regarded as atheistic materialism and linked to the idea of evolution, while Pasteur was a devoted Christian and was already opposed to such a hypothesis in principle. So he decided to use precise experimentation to prove that living beings could only come from other living beings, and not from any inanimate part of nature.

Pasteur and Pouchet both conducted a series of experiments to prove their hypotheses. In the end a special committee of the French Academy ruled in Pasteur's favor, further consolidating his fame. In the end, however, while reading through Pasteur's diaries, historians discovered that the scientist had skillfully concealed several weaknesses in his theory. If all of these difficulties had been revealed publicly, he would never have been declared the winner.

Lab diaries paint a different picture

The problem was that microorganisms exist that cannot be destroyed even by boiling water. Pouchet used hay for his experiments because he knew that it contained this very variety of highly resistant spores. Thus, he always reached the conclusion that even boiled hay can enable the spontaneous generation of life. Pasteur repeated these experiments himself and came to similar conclusions, but decided to conceal them. He preferred to use a solution of yeast which contained no such spores and enabled him to easily prove that spontaneous life could not occur.

If French society at the time had not been so inclined to favor the conservative ideas about the generation of life supported by Pasteur, the Pouchet-Pasteur debate would not have been decided so suddenly, as not all legitimate doubt had been dismissed. However, the public simply had the impression that it was a matter of objective facts, supported by carefully conducted experiments.

Today, Pasteur is regarded incontestably as a hero of nineteenth-century century science. Because of his talent for public display and his success in solve practical problems encountered in industry, agriculture and healthcare, he would certainly be the model example of a scientist even today, and would be eagerly supported by any country or company. Part of his appeal however did lie with the way his scientific findings were not socially confrontational, like those of Galileo, Copernicus and Darwin, but actually reinforced the ideological foundations on which society stood at the time.

LIFE

God's final message to his Creation: "We apologize for the inconvenience."

Douglas Adams (*So Long, and Thanks for All the Fish*)

What wiped out the dinosaurs?

When American geologist Walter Alvarez started studying rock formations near the Italian medieval town of Gubbio, Umbria, in the 1970s, he had no idea that he would uncover the cause of dinosaur extinction. At that time, Alvarez was developing a new method for determining the age of sedimentary rock. He chose the limestone strata by the old town in the middle of the Italian Peninsula because they offered the perfect spot for his experiments, since they comprised clearly layered rock which would not be difficult to date.

Many millions of years ago, when dinosaurs still roamed the planet, a large part of southern Europe was under water. The bottom of the sea is where sediments are deposited and in time become consolidated into rock. Through tectonic movements in the Earth, these submerged areas were later brought to the surface. The rocky region in the vicinity of Gubbio can thus reveal what was happening on the surface of the Earth 50 to 100 million years ago.

The Secret of Fossil-Free Layers

When Alvarez examined the sedimentary layers, he noticed a discrepantly darker line, a centimetre thick, which stood out from the surrounding strata. The older sediments under it were packed with the fossilised remains of small sea organisms from the Cretaceous period. Above the line, there were many less fossils to be seen, which is representative of the early Tertiary period. Yet surprisingly, the reddish clay layer, which separated both great geological eras, was completely empty of all fossils.

Alvarez knew that the borderline between the two periods corresponded with the extinction of the dinosaurs. Experts call it the "K-T Boundary" (from the traditional abbreviations for the Cretaceous and the Tertiary periods respectively) and it seems to indicate that a great catastrophe befell the planet. By studying old fossils, geologists were able to work out that half of all living beings were annihilated during that period, but they could only speculate as to the source of this cataclysm. Alvarez believed that analysis of this thin sedimentary stratum, which coincides with the K-T boundary and can be found all over the world – although it is thicker in America than in Europe – would uncover the true cause behind the 'Great Dying', as such global calamities are labelled, which wipe out more than half of all life on the planet.

The great disaster that struck Earth 65 million years ago was far from the only such occurrence in the planet's exciting history. Geologists have named several such events which precipitated mass extinction. A similar catastrophe happened 250 million years ago, at the end of the Permian period. Experts estimate that an unknown event suddenly obliterated 90% of all sea-living species and 70% of all land-based species. After this Great Dying, the cause of which remains unknown, but was probably connected to volcanic activity, dinosaurs took the planet by storm.

In the second half of the 1970s, Walter Alvarez joined forces with his father Luis Alvarez, a renowned Nobel Prize Laureate in Physics, and began studying the fossil-free borderline in order to shed some light on the extinction of the dinosaurs. The two scientists first wanted to find out how the centimetre thick layer came into existence, in the belief that this could point them towards the cause of the catastrophe. They decided to ascertain the speed with which the sediments formed by studying the deposits it contained of a rare element called iridium. Almost all iridium on the Earth's surface comes from the dust of small meteorites, which can usually be seen as shooting stars upon entering the atmosphere. When our planet was still forming, Earth's iridium was drawn to the core, which is why it is virtually non-existent on the surface. Because an almost equal quantity of meteor dust falls to the planet every year, geologists can determine how many years it took for a particular sedimentary layer to form by studying its iridium content.

Did a Mountain Crash into the Earth?

The scientists took a sample from the layer located near Gubbio, analysed it and discovered that it contained thirty times more iridium than normal. This was a clear sign that something unusual had been happening at the time from which the layer dated, something that could not be explained by conventional geological phenomena. To exclude the possibility of a local deviation, they performed the same analysis on a sample from the Danish cliff of Stevns Klint, in which they measured 160 times more iridium compared to the surrounding layers corresponding with the time before the K-T event.

Based on these results, they set out a daring hypothesis that the extinction of the dinosaurs 65 million years ago was brought about by a 10 km wide comet or asteroid crashing into the Earth. In their opinion, a mass the size of Mount Everest collided with our planet and caused such vast devastation that half of all life populating the surface at that time died out.

The impact of the asteroid or comet triggered giant tsunamis, caused countless fires and sent so much dust into the air that it took several years before sunrays touched the ground again. The lack of light prevented photosynthesis, which in turn meant that plants were unable to grow and food for animals grew scarcer. While food shortage was fatal for the dinosaurs, small mammals – distant relatives of humans – were able to survive.

In 1980, the father-and-son team published their findings in *Science* magazine, only to find themselves the target of heavy criticism. Other experts were convinced that the excess iridium was due to powerful volcanic eruptions and did not come from space. But eventually, the evidence supporting their hypothesis grew more substantial. Geologists discovered the same iridium-enriched layers in all parts of the planet. They also found specific types of rock which only form under extreme pressure, e.g. in a catastrophic collision.

A Hidden Crater in the Gulf of Mexico

In 1991, another great discovery was made and the meteorite hypothesis finally became an integral part of history textbooks. The remains of an enormous crater with a diameter of 180 km were discovered on the Yucatán Peninsula in Mexico. It was named the Chicxulub crater, after a village near its centre. The crater is one of the largest confirmed impact structures on Earth and its age corresponds with the reddish sedimentary rock layer. It is therefore in all likelihood the crash site of the asteroid, which wiped out the dinosaurs and cut short their 150-million-year-old reign.

The crater was first discovered several decades earlier, when geologists working for the Mexican national oil company Petróleos Mexicanos (also known as Pemex) spotted it in aerial footage from a survey of the surrounding region. They were not however allowed to discuss their discovery because of the confidentiality of corporate data. Only in 1981 did the company loosen its strict policies and allow geophysicist Glen Penfield to present the newest measurements at a scientific conference. As the turnout was rather poor, the lecture passed largely unnoticed. Penfield also failed to secure rock samples from drill holes made in the crater, which supposedly got lost.

A decade passed before Penfield co-operated with scientists searching for possible locations of an asteroid crater made some 65 million years ago. Pemex finally made the effort of finding the drill samples, which confirmed cause of the crater. The samples revealed rocks which can only form under tremendous pressure and in extraordinary conditions, such as when a mountain collides with a planet.

The afterlife of Henrietta Lacks

In February 1951, Henrietta Lacks, a thirty-year-old mother of five, went for an examination at the John Hopkins hospital after noticing blood stains on her underwear. Doctors were quick to conclude that she had most likely contracted cervical cancer. To confirm their diagnosis they took a sample of her tissue and had it analyzed. The news was anything but encouraging: the tumor contaminating the cervix turned out to be malignant. In October the same year, after only a few months' struggle with the disease, Henrietta Lacks died.

Immortal cells of a dying patient

Even though Henrietta and her family could not have realized this fifty years ago, her body's life had not ended yet. Some of Henrietta's cells are alive even to this very day, more than half a century after their owner's death. They are being carefully grown in numerous laboratories across the world and have even reached outer space after being sent into the orbit on a satellite, enabling researchers to observe the effect of zero gravity on cell growth. In fact, if we were to weigh all Henrietta's surviving cells together, they would probably largely exceed the weight of her entire body fifty years ago.

The fascinating journey of Henrietta's cells began in 1951 when a part of her tumor's tissue was sent to the laboratory of Otto Gey, a researcher at John Hopkins University who had been striving for decades to find a solution to growing human cells in a laboratory in order to facilitate cancer research. He struggled for a long time to find a solution, but cells simply refused to grow and divide in an artificially created environment. That was until he came across the cells from Henrietta's tumor and decided to try using them.

Miraculous HeLa cells

Even after several days in an artificial culture medium, Henrietta's tumor cells continued to divide successfully and there was no indication of them ageing and dying as samples had in all previous experiments. Gey named them after his patient: HeLa cells. He concealed the identity of his donor with great care and it was only revealed after his death.

However, as rapidly as HeLa cells multiplied in Gey's laboratory in 1951, they also spread inside Henrietta's body. Even though the patient died only half a year after her first visit to the hospital, the very cells that did her so much harm were soon helping to save many other lives.

In the 1950s a vaccine for polio was developed with the help of HeLa cells. They have gone on enabling an abundance of research up to the present and have led to the discovery of many important facts about life and the workings of the human body.

Immortal cell lines

Henrietta's cells were the first to be successfully grown in laboratory conditions and later used to create an "immortal cell line," the term used for cells which can – in an appropriate environment – continue to divide outside the body without aging. In normal circumstances, cells may only divide for a limited number of times (ten or more) before losing this ability and slowly dying.

Cell death after a certain number of divisions was in fact what stopped Gey from creating an immortal cell line from normal cells in the human body. It was only Henrietta's cells that displayed an ability to divide indefinitely. That was because they lacked the mechanism responsible for ceasing uncontrolled multiplication. Its inactivity failed to trigger cell death when HeLa cells had reached the normal threshold for duplication.

Cell aggression beats the competition

As HeLa cells turned to be very useful in numerous researches, Gey simply mailed them to colleagues all across the world so they too could grow them and use them in their work. Soon, other scientists reported that they too had created their own immortal lines of human cells. However, it soon turned out that these so-called new lines were really just a form of Henrietta's cells. They were found to be highly aggressive, capable of infecting other cell cultures and eventually eliminating the competition even when applied in extremely small doses.

In 1972, for example, Russian scientists sent their American colleagues six different lines grown in various parts of the former Soviet Union, but it was discovered that all of them were actually HeLa cells. Americans did not prove any more successful either: in 1968 34 lines were tested of which 24 were identified as Henrietta's cells.

Growing patient cells

As scientists perfected the methods of growing human cells and cells of other mammals in laboratories, they quickly began to wonder whether it was possible to create immortal lines for specific individuals as well. Organ transplantation always entails a risk that the immune system will reject a transplanted heart or liver. Doctors try to overcome this by prescribing special medications which transplant patients must take regularly for the rest of their lives and by seeking genetically compatible donors for organ recipients. A transplant is performed only if this match is close enough, since only then can drugs be of any use should the patient's system reject the organ.

Transplants using donated organs always entail large problems and high risks, so researchers started to think how they might grow healthy cells from a failing organ in a laboratory and reinsert them into the patient later on. This would prevent the risk of a rejection of foreign tissue, as the artificially grown cells would not differ in any way from those already in the patient.

Today, skin and bone marrow-growing therapies are already available. A sample of cells is taken from the patient and multiplied in sufficient amounts in a laboratory only to be transferred back into the patient later. However skin-growing can prove to be particularly difficult, taking several weeks to produce enough for transplantation, which might already be too late for those with heavily injuries.

Cells adapt to specific tasks

Another problem in growing cells from various tissues of the adult human body is that cells can only divide a limited number of times. Sooner or later the process comes to an end or the cells produced do not respond to our needs.

Even though the genetic code of every cell contains all the information it requires for development, most of the cells that make up an adult body have already lost the ability to use the entire code. In the environment of a specific organ, cells adapt to perform a specific task and this adaptation is usually irreversible.

When a cell takes on a function within a blood vessel, or in the heart, the skin or the nervous system of an individual it cannot "change its mind" and take on a different role. Scientists have not yet figured out what exactly happens within a cell when it adapts to a specific task, but the fact remains: a part of its genetic code is shut off for good.

The mother of all cells

However, not all the cells within the human body have this one-way limitation in terms of their function. Those that have the ability to develop into other types are called stem cells. Their main characteristic is the capacity they have of developing into more specialized cells. Bone marrow, for example, contains special stem cells which are capable of producing red blood cells to transport oxygen throughout our bodies. Similar stem cells that can produce other specific cells have also been found in various organs.

The capability to develop into any of several hundred cell types within the human body is possessed only by the very young stem cells that can be found in the first days following fertilization, in embryos made up of only some ten cells. These are called embryonic stem cells and only they have the rare property of developing into any cell type in different environments.

Immortal stem cell lines

Another important feature of embryonic stem cells is that they can be used to create a cell line. If grown correctly, they can multiply indefinitely without growing old and dying after some ten divisions, as is usually the case with already differentiated cells such as skin or bone marrow cells. The first line of human embryonic stem cells was created in 1998 in the US.

Today, several hundred exist in the world. In the US, however, following President George W. Bush's decree, public research funds can only be used on the 22 lines created before the beginning of his mandate in 2001. This issue is once again gaining in importance since the president vetoed the Act of Congress that would have approved the continuation of federal funding for this important and vital scientific field.

A Russian Indiana Jones

It is no secret that the democracy of a given country hardly necessarily reflects the support given to science by its authorities. It has often been the case that totalitarian countries were more generous towards science than their democratic neighbors. Of course, this observation only holds true as long as scientists do not interfere with government policy, and only if their achievements contribute to boosting the country's reputation. If a scientist succeeds in obtaining the image of a "national hero", any government, especially a totalitarian one, will love him.

A young visionary

When Russia was immersed in revolutionary turmoil during both World Wars, Russian genetics and agricultural science were among the most advanced in the world. The greater part of the success of Russian life sciences in the first half of the 20th century may be attributed to a young, talented, and hardworking agronomist and outstanding organizer, Nikolai Ivanovich Vavilov.

Nikolai was born in Moscow in 1887 as the oldest of four children in a wealthy merchant family. His younger brother Sergey became a renowned physicist and the president of the Russian Academy of Sciences. Nikolai himself had been interested in science - plants in particular - ever since he was a young boy. He graduated at the Moscow Agricultural Institute in 1911 and began his scientific career in the field of applied botany or cultivated plants. The crucial moment in his scientific development came in 1913 when his superiors decided to send him to England to further his knowledge in genetics. William Bateson, for whom Nikolai worked for a while, was at the time the world's foremost geneticist and had a great influence on the young scientist. Nikolai returned to Russia shortly after the outbreak of the First World War, swiftly finished his post-graduate studies and became a professor of agronomy, botany and genetics at the Saratov University.

Soon after the Russian Revolution of 1917, Lenin appointed him chairman of the Institute of Applied Botanics in Leningrad which, under his leadership, quickly became the world's foremost institution in the field of cultivated plants research. At the peak of his career around 1934, Nikolai employed 20.000 assistants from all across the Soviet Union. Surprisingly, neither Nikolai nor his younger brother Sergey ever joined the Communist Party. They did, however, support the social and economic goals of the revolutionary regime.

The largest gene bank for plants

Nikolai Vavilov was incredibly energetic and hardworking. According to accounts given by his colleagues, he slept only a couple of hours a day, requiring his technicians to be in the field at four o'clock in the morning, ready to start work. As he understood the basics of genetics, which he acquired during his studies in Europe, he knew that crossbreeding would enable him to create better and more disease-resistant plants, as long as he had enough different species of plants at his disposal.

The idea was simple: in order to grow the best plants for producing food he needed to obtain a larger assortment of plant species so that he could crossbreed and discover those which were the most appropriate for cultivation. The more different varieties of wheat, rice and potato from different parts of the world he had, the easier it was to grow new sorts with higher immunity.

Of course, he did not have a collection of different varieties of plant seeds at that particular time, so he took advantage of his position at the Institute of Applied Botanics and organized more than one hundred scientific expeditions to Asia, Africa, America and the Middle East, gathering samples of more than ten thousand plant species. Vavilov traveled to distant mountains, forests and meadows, sending rare sorts of food plants, such as rice, wheat, corn, barley, oats and potato back to his institute in Leningrad. Naturally, he was also very interested in hitherto lesser known crop plants such as lentils, chickpeas, soybeans and several others. He created the largest seed bank in the world, containing more than two hundred thousand samples, thirty thousand of which alone were sorts of wheat. Some of the plants he discovered were still unknown in Europe at that time.

Plant hunting

Sometimes, Vavilov could acquire his plant seeds from markets in remote places with relative ease; at other times, however, he encountered greater difficulty. One such occasion was when he barely managed to return alive from one of his African expeditions to Ethiopia (then Abyssinia) in the 1920s. In his writings he describes how he had to paddle his way among threatening giant crocodiles of the Blue Nile, searching for a rare sort of wheat and barley. One night he was attacked in his tent on the river bank by a swarm of spiders and scorpions. These he succeeded in tricking back outside by lighting an oil lamp in front of the tent and so luring the dangerous creatures back out. Even deeper in the African wilderness he was taken prisoner by armed bandits whom he later evaded by means of a proven Russian method: he got his captors drunk on some fine vodka and escaped when they fell into an inebriated sleep.

In 1923 he built more than one hundred experiment stations for testing crop plants. A few years later, he united all of his research centers under the Lenin Academy of Agricultural Sciences and became its president. During the same year he was elected a life member of the Soviet Academy of Sciences.

In 1926 he published his famous treatise on the geographical origin of cultivated plants. He argued that every existing "domesticated" plant has its particular place of origin somewhere on the planet where it is also possible to find the greatest genetic diversity of this individual species. If, for example, we were searching for new sorts of potato, we would do best to look for them in South America because potato originates in the area of today's Peru, Ecuador and Bolivia.

The quarrel with Stalin's protégé Lisenko

When Nikolai was in charge of the Institute, Russian plant geneticists were among the best in the word. He openly supported international cooperation and the transmission of new knowledge from the West. This was of great benefit to Russian science, but did not comply with the regime's ideology. When Stalin's agricultural reform with kolkhoz farms (Soviet collective farms) failed to have the required success, the regime needed a scapegoat. Nikolai Ivanovic Vavilov was a suitable victim as he had already been exposed to criticism on account of his expensive and frequent journeys to far corners of the earth. Vavilov tried to defend himself, but his arguments did not help.

The greater part of the most important positions in the field of Soviet biological and agricultural research were appointed to the charismatic yet poorly educated agronomist Trofim Denisovich Lysenko, who caught the government's attention with his alternative approach to biology. He argued for a sort of Lamarckism. In his opinion, Vavilov's method of searching for new species of plants by crossbreeding was, academically speaking, excessively abstract and bourgeois. He suggested a far simpler and in his view more effective method of "training" plants, to "teach" them how to adapt the necessary form and structure that would enable them to produce the maximum possible amount of food.

Lysenko personified the mythical ideal of the Soviet country-boy genius, who offered Stalin's regime heroic achievements of "genuine home-made" science. When he was awarded the Order of Lenin, the Pravda newspaper published a letter of thanks from his parents, addressed to Stalin himself:

> The day we found out that our Trofim was awarded the Order of Lenin was the happiest day of our lives. How could we have ever even dreamed of such a great honor bestowed on us, poor farmers from the village of Karlovka?

Lysenko's "agricultural miracles" did not, however, work in practice. The greatest damage his lack of knowledge caused his country's economy was inflicted he nearly destroyed the Soviet dairy industry. By crossbreeding pedigree cows with common bulls he managed to ruin the achievements of several generations of elaborate cattle breeding.

Starvation in a warehouse of food

Moments away from the Second World War, Stalin's purges finally caught up with Nikolai. On August 4th, 1940 he was arrested on a trip to the Ukraine and charged with sabotage and collaboration with bourgeois reactionary forces. At first, he was sentenced to death, but the punishment was later commuted to ten years of imprisonment after his brother Sergey intervened. When he became a member of the distinguished British Royal Society in 1942, the Russian secret police reopened his case, but they were too late. Even though Vavilov had dedicated his life to studying agrarian sources of food, he died of starvation in his cell on January 16th, 1943.

There is an anecdote from the siege of Leningrad during the Second World War which attests to Vavilov's personal charisma and influence. When people were dying of hunger on the streets of the city, a few of his colleagues selflessly guarded the Institute with the priceless collection of seeds they had gathered with great effort from remote parts of the world. At least ten of his colleagues succumbed to starvation, without even touching the valuable samples of potato, rice, corn and other crops from faraway places.

Creating a second paradise

Born three hundred years ago, the Swedish naturalist Carl von Linné became one of the most notorious figures on the European scientific scene. He had a very high opinion of himself, as is evident from the autobiography in which he spared himself no applause for his achievements: "There has been no greater botanist or zoologist. Nobody has written more books, in such detail, as systematically, and on the basis of his own experience. Nobody has revolutionized science so thoroughly to begin a new era." However, he was not alone in singing his praises, as many thinkers of the time recognized his greatness. Jean-Jacques Rousseau sent him a message that read: "Tell him I know no greater man on earth." And Johann Wolfgang von Goethe wrote: "With the exception of Shakespeare and Spinoza, I know no one among the no longer living who has influenced me more strongly."

Becoming the second Adam

The 18th century is often referred to as "the age of classification" and Linné was its most typical representative. Even after three centuries, Linné is remembered mostly because he created the efficient and simple system of nomenclature and classification for living beings which is still used today. Systematization was not, though, his only scientific passion. He also had the visionary idea of recreating paradise on Earth in the university town in Sweden where he lived. He wanted to gather plants and animals from all over the world and breed them in such a way that they could survive the relatively unfriendly Swedish climate.

Linné saw himself as a second Adam. In paradise the biblical Adam knew the names of all the animals put there by God. Linné wished to recreate this natural paradise in the botanical gardens of Sweden's Uppsala. He thought that the original biblical paradise was on one of the smaller tropical islands from which plants and animals spread across the rest of the world, adapting to other, colder climatic conditions, but remaining more or less unchanged. His "scientific" idea was that this otherwise spontaneous adaptation of originally tropical plants could be artificially encouraged. He was certain that if the plants were able to adapt to a different climate before, they could be made to do so again.

A new heaven on Earth

His mission was to create a new heaven on Earth in Sweden. He sent his assistants and students to all the faraway places of the world so they would bring him exotic plants, which he would then try to adapt to the unfavorable Swedish climate. He first planted the specimens from warmer places in the southern part of the country, then gradually moved them toward Uppsala. From China, for example, his envoys tried to transfer tea plants and rice as well as mulberries on which silkworms could be bred.

Linné's idea of growing as many plants as possible in Sweden was driven mostly by his desire to make Sweden an economically independent country. At the time, economy was interpreted in a very mythical way. All of a country's food supply problems would be solved, it was believed, if scientists succeeded in creating paradise-like conditions, an ecologic harmony that God had originally created in paradise and nowhere else. Some scholars claimed that God had spread the treasures of nature across different parts of the world to encourage international trade, but Linné did not agree.

However, the megalomaniac scheme of the visionary naturalist hit an insurmountable obstacle, the Swedish weather. Despite his cunning attempts to adapt exotic plants gradually to the local climate, only a few managed to survive a longer period in their new environment. It is interesting that Linné almost made an important breakthrough with the calorie-rich potato, but he was unfortunately convinced that it was poisonous. It was supposed to be related to the very poisonous deadly nightshade (*atropa belladonna*), so he was certain that even pigs would not touch it. Sadly nobody hit upon the idea of cooking only the plant's bulbs and eating them as we eat potato today. Nevertheless, they had already discovered that potato could be used to make liquor and even powder for wigs, which were at the time still very fashionable. The author of this idea actually became the first woman to be accepted into the Swedish Academy of Science and the only one who received this honor before the beginning of the 20th century.

Names and surnames for plants

While gathering exotic plants all over the world, Linné's emissaries introduced local scientists to his system for classifying living beings. This was quickly accepted on account of its simplicity, becoming much more successful than his idea of duplicating paradise in Sweden.

Linné's system is very effective because it describes any plant or animal with only two Latin names, which is similar to the way people identify themselves with a name and a surname. He chose Latin names to make the system universal and prevent some nations from feeling left out. Before, naturalists used different and often highly impractical methods to name nature. It was quite common that a plant would get a descriptive name which was fifteen or more words long.

It is not so well known that his system of plant classification was very controversial when it was first published. His method of classification was criticized as being unnatural, a system that he had chosen to force and divide plants into imaginary groups. Another aspect of his method presenting a great problem for his critics was his decision to classify the plants according to their reproductive organs.

This might seem strange today, but it was not until the end of the 17th century that naturalists discovered that plants also reproduce sexually. Linné's principle of classifying plant species was based on "plant marriages", as he explicitly described his method. He introduced a system of classification derived from the number of males (stamens) and females (pistils) present in a particular plant marriage.

An immoral classification of plants?

In his fundamental work *Systema Naturae,* first published in 1735, Linné began by dividing the kingdom of plants into classes according to the number, size and position of the male members of the plant marriage, today called stamens. These classes were then further divided into orders depending on the number, relative size and position of the female members of the marriage or pistils. Based on the structure of their flowers or fruits these orders were then divided into genera which, in turn, were divided into species according to the leaf structure or other characteristics of the plants.

Linné's system of classification is of course an abstract creation. Even though it was founded on the structure of vegetal reproductive organs, it was not derived from actual knowledge of the way these organs function, but aimed at sorting plants into an efficient system and naming them. It was based only on the structure of the reproductive organs and had nothing to do with their function.

Of course, the guardians of public morals of the time wanted to protect young girls from the bad influences of an education in botany. Linné spoke of the principal organs of plants as of a man and a woman on their wedding night and used other suggestive names such as blanket and bride's bed.

In the tenth edition of his book *Systema Naturae* he introduced a new term Mammalia which comprised the large group of animals that we still call mammals. It is interesting that the common characteristic he chose to name this group after were woman's breasts, even though he could have easily decided for a different trait typical of this group. In the same edition of the book he also introduced the term *Homo sapiens* referring to mankind.

Alexander von Humboldt - adventurer and scientist

There was a time when adventurers did not venture to remote places of the globe and conquer high mountains and peaks only to get an adrenaline rush and test the limits of their capabilities, but were driven by a desire to discover the unknown. The principal figure among these adventurers who were at the same time brilliant scientists was undoubtedly the German scholar Alexander von Humboldt, whose name became a synonym for the true naturalist. Even though he is hardly ever mentioned today, in his time he enjoyed a superstar reputation similar to that of Albert Einstein in the twentieth century. At the peak of his popularity, contemporary sources testify that only Napoleon was more famous.

The life story of Alexander von Humboldt is interesting in itself. He and his brother Wilhelm, who was also a very successful scientist and politician, grew up in a wealthy Berlin family. Their father was an officer in the Prussian Army and their mother was from a rich, noble family, so they were well provided for. After the early death of their father, their mother hired the best private tutors to ensure that her sons received the best education possible at the time.

Their mother wished for her sons to pursue illustrious careers as statesmen, but her death in 1796 liberated the young men from the pressure of fulfilling her ambitions. Alexander, the younger of the two, had always been enchanted by accounts of travels to distant places, so he soon decided to join one of these expeditions himself. As he had inherited a fortune, he had no difficulty finding the means to pay for his journey. All that needed to be done then was to organize the expedition, and this, more than two centuries ago, was not an altogether easy task.

The shocking experiments of a mining inspector

To learn as much as possible about such expeditions he made the acquaintance of some of the scientists who had, among others, accompanied James Cook on his legendary voyages. He also traveled across Europe to meet a number of scholars and make important contacts. He visited several renowned universities where he studied everything from economy and languages to astronomy, biology and geology with the most eminent professors of the time.

Although this might sound strange today, geology at the time was the discipline in which the most radical discoveries were being made, attracting the most interesting scientific debates. Alexander finished his formal education at twenty-three years of age at a mining academy and became a mining inspector. During the years he spent in the mines he introduced many improvements in the working conditions, especially in the area of safety, so he was very popular with the miners.

All along he was very interested in science and conducted some experiments himself. As a geologist he was fascinated by the Earth's magnetic field and the similar phenomenon of "animal magnetism" which was being researched by other scientists such as Luigi Galvani, who studied the effect of electricity on frog's legs. Humboldt was so impressed with this new branch of science that not only conducted electricity and magnetism experiments on rocks and frogs, but on his own body as well.

His hypothesis was that metal electrodes did not cause the contraction of muscles as an external source, but merely stimulated their inner properties. He first tested this hypothesis on plants and animals, then on himself as well. The descriptions of experiments which he published in 1777 in a notorious book were in places quite frightening. He used various electrodes which he attached to different positions on his back to trigger muscles and evaluate the intensity of the pain that spread across his back. One time, the back injuries he inflicted on himself were so severe that the attending doctor insisted on stopping the experiment, but Humboldt kept going on, driven by his passion for new experience and knowledge. His notes even contain some bizarre statements such as: "The frogs on my back then started to hop."

Despite his extensive experimentation he was very surprised by Alessandro Volta's discovery that making a battery does not require the use of animal tissue. He felt disappointed because, despite all the extremely painful experiments he had endured, he had been unable to reach this conclusion, which quickly made Volta famous. It was at this time that he stopped working on electricity and decided to dedicate himself to botany.

The most important years of Alexander's life passed during the transition into the nineteenth century. It was in the period between 1799 and 1804 that he and the French botanist Aimé Bonplandon traveled across a great part of what was then a completely "wild" South America. Together they walked across the territories of today's Venezuela, Columbia, Ecuador, Peru, Mexico and Cuba. They concluded their journey in Washington where they met Thomas Jefferson.

Conquering the supposedly highest mountain on earth

During his travels, Alexander continually sent letters to his friends and family, containing interesting accounts which were enthusiastically summarized by European newspapers, allowing the public to follow his exploits a couple of months after they took place. When he returned to Europe from his journeys, he brought more than 60,000 samples of plants and a number of animals that had until that time been unknown. However, his most celebrated feat was his ascent of the volcano Chimborazo, the tale of which dazzled the European public.

At the time it was believed that this South American volcano, which was called the Snow Mountain by the natives, was the highest mountain in the world. However, this ascent was not only a great adventure, it was also a serious scientific expedition. Humboldt's monograph on the mountain, which he published on his return home, set the foundations for many other similar scientific works. In his book Humboldt gave a detailed description of the geography as well as of the plant and animal life on the mountain, focusing on how the change in environment at different heights influenced the types of flora and fauna that could exist there.

At the time it was still unclear if a human being could even survive venturing to such heights. But Humboldt was in great physical shape and, being used to experimenting on his own body, was not afraid of the ascent. His reports of the climb were actually what finally made him a real celebrity. Avid readers were impressed with Alexander's "mountaineering" as much as his scientific achievements.

At the beginning the way up was easy, but gradually the terrain started to become more difficult. Unfortunately, not every member of the expedition had suitable equipment, and certainly not the kit which even allows people with no mountaineering experience to reach the summit today. From the description of the difficulties the adventurers experienced during the ascent it is evident that they were affected by altitude sickness and, in addition, lacked proper clothing to protect them from the cold. Nonetheless, they continued to diligently measure the air pressure and temperature and carefully observe their surroundings all the way to the top. Despite all obstacles Humboldt was determined to persevere and rise as high as possible. At first, the fog prevented them from seeing the summit, but then the weather cleared up and their goal no longer seemed to be as far away as before. However, they came upon a hollow which, with the equipment they had, they were unable to cross.

The ice-clad volcano was a tempting goal, but the combination of cold, snow, lack of oxygen and other completely objective obstacles prevented the scientific expedition from conquering what was believed to be the "top of the world". Even though Humboldt, Bonpland and a local guide did not succeed in reaching the summit of the 6268 meters high mountain, the 5875 meters meant an altitude record, as until that time nobody was recorded to have climbed higher. Another witness to the difficulty of the ascent is also the fact that the peak of Chimboraza was not conquered until 1880, almost a hundred years later.

After his return to Europe, Humboldt wrote down everything that he had discovered and learned during his five-year journey. Over the following twenty years he wrote a number of books and discussions which attracted many enthusiastic readers. One of the most famous admirers of Alexander von Humboldt was Charles Darwin. He actually took the accounts of Humboldt's travels with him on his famous voyage on the board of the Beagle and used it as a sort of a guide through the poorly explored territories of South America. When he reached Brazil he made an entry in his diary: "I have always admired him; now I worship him."

Feral children

In 1996, Ivan Mishukov left home. Even though he was only four years old, he felt that the situation at home was far too chaotic for him to bear. He simply could no longer endure in this environment. His mother was incapable of taking care of him and he often got into trouble with her drunken lover, so he ran away and decided to try and survive on the street. Apart from homeless people, a great number of stray dogs then roamed the streets of Moscow, animals that the inhabitants of the city could no longer care for or had simply got tired of and let run free. These dogs gradually grew wilder and started coming together in packs.

While begging on the street, Ivan quickly aroused the compassion of the passers-by because of his very tender age, and since he managed to gather a fair amount of food every day decided to share his spoils with one of the groups of stray dogs. He gradually gained their trust and eventually became accepted as the leader of the pack. Ivan and the dogs lived in a sort of symbiosis. The child would bring them food, which he had accumulated by begging, and the dogs would give him protection and keep him warm at nights, especially during the cold Moscow winters.

Once the police found out about the way the boy lived, they naturally attempted to capture him, but the dogs stopped them from doing so on several occasions by protecting him so he could escape. It was only after they lured the dogs away with food and isolated the boy that they could get to him. He had been on the streets for two whole years.

Because Ivan had spent the first four years of his life among people, he could talk well enough, so he was able to reenter human society quickly after only a short period in a children's shelter. He soon started going to school, found a home with a foster family and has since been living the relatively normal life of an average Russian child.

All "natural" children

Ivan is just one of a number of children known to have spent a part of their childhood living outside human society in total isolation or in the company of wild animals. These children have been known as "feral children". The semi-mythical, fairytale-like stories about how animals took care of them and raised them have been present throughout history. During the Enlightenment there was a surge of interest in these "all natural children" among the scholars of the time. Thinkers were very interested in learning what separates man from animal and wild children, as a sort of borderline case, were therefore fascinating.

In reality feral children were completely normal young people who had been growing up isolated from human society, had not adopted the customs characteristic of its culture and also for the most part could not speak. It was the question of whether such children could be taught to speak after they had passed the usual age at which infants begin speaking that became the central subject of the scientists' interests whenever they came into contact with a feral child.

About two centuries ago, the scientists of the time were fascinated by two boys in particular who suddenly found themselves at the center of everyone's attention. The first boy was retrieved in 1800 in Aveyron, France, having regularly been seen running naked through the woods for several years, and always managing to get away. He became known as Victor of Aveyron and after his return into civilization spent several years in the care of one Doctor Itard. The doctor examined him carefully and tried to teach him how to speak, but failed to do so despite years of effort. Victor's case was revived in the 1970s in François Truffaut's famous movie entitled *L'Enfant sauvage* (The Wild Child) in which the director himself played the role of Doctor Itard.

The other case was perhaps even more mysterious. On Monday, May 26th 1828, the streets of Nürenberg were calm and empty; it was a public holiday. Suddenly, at four o'clock in the afternoon, an unusual boy was spotted wandering around aimlessly. At first people thought he was lost, but when he started repeating the sentence *I want to be a cavalryman, as my father was*, the meaning of which he could not understand, it became clear that this was no ordinary boy. They tried getting him to tell them where he had come from, but did not succeed in learning anything meaningful. However, they were surprised to find out that he could write his name: Kaspar Hauser.

The boy quickly became the attraction of the Europe of the time. All the newspapers wrote about him, and many famous people wanted to see him. Hauser also began to receive intensive lessons which turned out to be of additional benefit to history as he learned to write, so he could later use his slightly broken German to tell the frightening story of his life as he remembered it. For more than ten years he had been locked away in a small basement space. In it there was a pile of straw on which he could sleep, a woolen blanket and two wooden toy horses. Each day, while he was sleeping, an unknown man would usually pay him a visit and leave some food and water in the tiny room. This monotonous hermit's life lasted until a few weeks before he was released when the man started to teach him how to write his name and pronounce some sentences, which he repeated when he was found.

The girl from the American suburbia

A similar ordeal to Kaspar Hauser's, at the beginning of the 19th century, was experienced more than a hundred years later by an American girl called Susan Wiley (named Genie in academic literature on her case) from the suburbs of Los Angeles. In

1970 she caught the attention of the employees of a social center where she and her almost blind mother sought refuge. She appeared to be about eight years old, but was actually already thirteen. Later it turned out that, like Kaspar Hauser, she had spent most of her life locked up in a remote room of the family house, isolated from the sounds of the street, only hearing her family when they yelled.

Her suffering was caused by her apparently disturbed father, who would for several years sleep in the armchair of his living-room clutching a loaded rifle in his hands and "guarding" his house. The girl's mother had lost most of her eyesight and was completely dependent on her psychotic husband, who was convinced, in a perverted way, that he could protect his daughter from the corruption of the outside world by hiding her away. Later, when her father was already on trial for the mistreatment of his daughter, he seemingly realized the wrong he had done and committed suicide.

Soon after she was liberated, Genie, much like Viktor and Kaspar, found herself surrounded by scientists. After some deliberation, the group of theorists and therapists responsible for her well-being decided that it would be best if the girl were to move in with one of their families. During the next few years she tried to get used to life in human society and these proved to be the happiest time of her life. She was also frequently visited by a young postgraduate student called Susan Curtis who taught her how to speak. It was the very question whether they could teach her to speak at thirteen years of age that intrigued the scientists the most, and was also the subject of the student's doctoral thesis. Genie became quite successful in picking up new words, but could never learn how to form longer sentences. She had no problems with the vocabulary, but could never learn the rules of grammar.

Chomsky's theory

The seventies were also the time of the growing popularity of a theory advanced by Noam Chomsky, according to which a set of universal grammatical rules was supposedly inherent and biologically determined within the structure of the human brain. According to Chomsky, when a child is exposed to a language environment this natural ability, inherent to all people, simply adapts to the language in which the child is being raised. That is why the period during which a child can learn a language spontaneously is believed to be limited. The upper limit was estimated to be right around thirteen, so Genie was an ideal test subject for confirming the hypothesis of the existence of a critical period in the development of a child.

Unfortunately, scientists did not reach any other important conclusions about language from their experience with Genie. Some doctors argued that it is not entirely clear whether Genie might have been born with brain damage or sustained an injury later, which could explain her severely limited ability to form sentences. Another question was also whether she had been taught long enough to learn all the basics of the language. Nevertheless, a great number of books and scientific discussions were written about her case.

After the four years she had spent living with one of the scientists' family, she had to return to her mother, because the researchers did not receive further financial support for their project. After only a couple of months, though, it became clear that her blind mother was unable to take care of her. They tried to find her a surrogate family, but she could not adapt to any new environment and thus was never allowed to stay for more than a short time. After several failed attempts to find a home, when because of bad behavior and intentional vomiting she had even been badly beaten by one of the foster families, her mental health deteriorated substantially. She withdrew completely and stopped talking. Today she lives in an adult foster care home and her condition is supposedly much worse than it was only a couple years after her liberation when she was still at the center of scientific attention.

The war of images

1906 was the first time a Nobel Prize was awarded to two scientists. The prize for medicine was shared by the Spanish Santiago Ramon y Cajal and the Italian Camillo Golgi who earned it for their groundbreaking research on the structure of the nervous system. Although they worked in the same field of science, the first time they ever met was at the award ceremony in Stockholm.

Traditionally, each laureate is given an honorary lecture in which he can describe his work in detail. Golgi's lecture was one day before Cajal's and some controversy between the two was to be expected. Even though both were doing research on the anatomy of nerves, each of them had his own theory on how the nervous system was built and how it functioned. Despite the fact that both of the scientists were highly accomplished professors, one could sense a strongly skeptical attitude towards some of the main conclusions each supported during their presentations.

As a child, he blew up his neighbor's door with a cannon

In the early years of his life, Santiago Ramon y Cajal showed little promise of becoming a distinguished scientist. He was a very troublesome child and was expelled from several schools for his low grades and misbehavior. Once, when he was eleven, he even ended up in jail: During summer holidays, he and his gang, of which he was also the leader, had constructed a live cannon from scrap metal. Naturally, the youngsters had to test their artillery piece and the neighbor's new garden door was the perfect target artillery piece.

After the explosion, which was devastating, the neighbor complained to the mayor who sent an officer to bring the eleven-year-old Santiago into custody. Of course, his father was furious and insisted on making this outrage his son's final lesson. Santiago had to spend four days inside a smelly, dirty cell on nothing but bread and water. His mother persuaded the guard to let her sneak in some food, but her son still had to spend those nights and days in solitude.

The severe punishment was obviously not enough to make him mend his ways, however. He and his gang constructed yet another cannon, only for it to explode while it was being tested. They continued experimenting with other ballistic methods which could have very easily ended with a tragic result. Once, the barrel filled with gunpowder exploded close to Santiago's face, but luckily did no more harm than leave him with an eye infection and a permanent scar on his iris.

However, Cajal was not only technically inclined, but also had great interest in drawing. In fact, he wanted to become a painter. His father, a university professor of anatomy, fused his son's love of drawing with his enthusiasm for anatomy and got him interested in biology. The two supposedly stole corpses from a nearby cemetery then dissected them, and the youngster used his talent for drawing to learn the technique of anatomic illustration.

In his memoirs, Cajal singled out this unusual period of his life as contributing the most to his later decision to become a scientist: "In front of the great anatomic desk which covered the dissecting table the brain as well as the stomach first shriveled in disgust. But they soon became used to it, and the corpses did not lead me to sad thoughts anymore, but reminded me of wonderful creations of life."

After finishing medical school he joined the Spanish Army as a doctor for a couple of years and spent a year on Cuba. There, he was unfortunate enough to contract malaria as well as tuberculosis, but this did not prevent him from marrying and having seven children after he returned. In 1881 he became a professor in Valencia, but it was several years earlier that he had used his modest savings to buy an old microscope with which he studied the structure of tissues and similar biological preparations.

A method successful because it rarely works

In Madrid in the year of 1887, when he was thirty-five, he met a psychiatrist friend who had just come back from Paris and brought with him a sample of brain tissue, prepared according to a special method invented by Camillo Golgi fourteen years earlier. Cajal was at that time compiling a book on the techniques used in histological research, which he intended to illustrate himself. He had a lot of difficulties with the studies of the nerve tissue, so his introduction to Golgi's new method was a true revelation.

The essential advantage of Golgi's neuron staining method was that it almost never worked. It only colored one in approximately a thousand neurons, which was extremely important for observing the structure and functioning of the nerve tissue. This made it possible to observe a single neuron in a mass of neurons and examine it in detail. This is something like having a bowl of pasta in which most of the spaghetti are transparent and invisible and one or two are dyed in dark yet vivid colors.

After learning about the new "black reaction" method, as Golgi's staining technique was called, Cajal moved the focus of his research and came to many key conclusions about the structure and functioning of the nervous system. Before, it was believed that the brain was a mass of intertwined connections constituting networks, while Cajal, with his anatomical studies, demonstrated that the nervous system was also made of individual cells which were called neurons and through which nerve signals traveled.

Even though both scientists used the same staining technique that Golgi had developed, their conclusions were diametrically opposed. The principal dispute was over whether neurons were completely independent and separate cells or whether they formed some sort of homogenous network.

The objectivity argument

When it came to interpreting their argument, drawings were of key importance. The difference between Cajal and Golgi was their differing opinions on how a scientist should ensure that his data is as reliable as possible. Basically, they were waging something like a war of images. They both criticized each other for not being objective: Cajal defended an unaltered representation and accused Golgi of intentionally interfering and modifying his descriptions, so that they corresponded to his own theoretical preferences.

Cajal was convinced that a scientist should only copy what he sees in as much detail as possible and try not to interfere with that in the image he produces. According to him, a scientist should be like a camera, only transferring onto the image what he sees under the microscope. Golgi, on the other hand, was of the opinion that a scientific drawing should demonstrate the essence of the phenomenon it is describing and try to understand it. The duty of the scientist was supposed to lie in modifying a drawing, so that it reflects an ideal example, even though it might be composed of several examples of what can be seen under the microscope.

Lorraine Daston, a director of the Max Planck Institute for the History of Science in Berlin, and Peter Galison, a professor of art history at Harvard University have emphasized in their book *Objectivity* (Zone books, MIT Press, 2007) that in trying to understand this interesting episode from the history of science it is necessary to realize that both scientists argued in good faith and that their sincere personal integrity should be viewed as the ideal of the true scientific approach. Both were convinced that their methods and consequently their conclusions were in accordance with the strictest principles of scientific work, so they firmly stuck to their beliefs.

As the two historians elaborate their book, scientific objectivity has a checkered history. The scientists of the Age of Enlightenment, for example, gradually felt obligated to improve their drawings of plants and animals and ended up creating much better and more beautiful images than those they could observe in the nature. Golgi, who followed this enlightened ideal, made his drawings of the nervous system illustrative by allowing them to be interpretative, because he thought that best met the goal of his inquiry.

However, as the historians clearly demonstrate with numerous examples, some scientists began to look upon such practice as a sin. Around the middle of the nineteenth century a new ideal of pursuing objectivity appeared, one which was in time defended by Cajal. "Let nature speak for itself!" became the motto for understanding scientific activity and the main question became how to describe and represent the world without creating the feeling of the presence of an actual observer. It was because of the pursuit of different ideals of how to portray objectivity in science that the two Nobel laureates began their dispute and their so-called war of images.

Hobbits from the isle of Flores

A decade ago on the Indonesian island of Flores, archeologists discovered the remains of primitive stone tools made by our distant ancestors. In itself, the find would have been nothing special if analysis of the sharp-cut stones had not revealed that the artifacts were at least 700,000 years old. They dated from a time when this area was believed to have been uninhabited.

The Indonesian island of Flores lies north of Australia between Timor and Java. Even though it seems to be near Southeast Asia, it lies on the other side of the Wallace Line which separates the zoogeographical regions of Asia and Australia. The line represents a natural biological border which, because of strong ocean currents, most animals simply cannot cross. It was first described by the nineteenth-century naturalist Alfred Russel Wallace, who independently reached the idea of evolution by natural selection at about the same time as Darwin. Because of the geographical separation, the evolutionary developments of species on either side of the Wallace Line were different, gradually causing species which were once the same to differentiate.

The discovery of a small-skulled dwarf

Before the Stone Age tools mentioned above were discovered, it was believed that people had only crossed this natural border and arrived on Flores several tens of thousands years ago, much later than the age of the stone tools were made and used found. Australian professor of archeology Michael Morwood, who first examined the stone tools, quickly decided that the find deserved to be studied more thoroughly. He and his colleagues estimated that the Liang Bua cave ("cold cave" in the local language) on Flores was a place where it might be possible to find remains of prehistoric men. Their excavations of the cave proved to be technically very demanding. This was because they had to dig another cave, more than fifteen meters deep, in order to reach the layers of sediments many tens of thousands of years old.

After digging for two years, carefully sifting and searching through everything they brought to the surface, they finally had some luck. They found a tiny arm bone which, at first, they could not even identify. During the following year of digging the only human remain they found was a tooth. In 2003, when their funds for further excavation began running out, they finally stumbled upon a well-preserved human skeleton. Because the skull was very small, they were first convinced that it had belonged to a little girl; but they soon noticed that it also contained a wisdom tooth, which little children do not yet have. As it turned out, they were not dealing with a child, but with a humanlike dwarf creature of the female sex, around 30 years old, that had an extremely small head. From her bones they estimated this midget was no more than a meter high.

Because of the small skull and other features that they were able to discern from the shape of the bones, the finding was classified as a new, until then still unknown biological species named *Homo floresiensis*, and the skeleton of the discovered dwarf dubbed LB1. On the same site they later found the remains of twelve other skeletons, all belonging to this extinct species of prehistoric dwarf. Because of the popularity of the Lord of the Rings film trilogy at the time, the media named this supposedly new species of prehistoric human "hobbits".

In the same layers, the remains of Stone Age tools, charcoal and elephant bones were also found which meant that these so-called hobbits had used fire and hunted elephants, making the find even more unusual. The discovered skeletal remains bore witness to the fact that hobbits had even smaller brains than today's chimpanzees. Their brain volume amounted to a mere third of the average human brain today, but at the same time these hobbits were supposedly capable of using fire and crafting tools.

The surprises kept on coming. By dating the layer of sediments in which the bones were discovered, archeologists found out that the sediments were no older than 30,000 years. After applying an even more precise method of measurement, which makes use of the ratio of carbon isotopes in the bones, they concluded that the discovered hobbit bones had to be between 13,000 and 94,000 years old. This meant that these dwarves had for a time lived during the same period as the modern man.

Extraordinarily developed humans or hobbits?

The story of the discovery took another twist when, in December 2004, the "hobbit" bones were temporarily passed into the hands of a 79-year-old expert on Indonesian paleontology, Professor Teuku Jacob. After several months of research, he came to the conclusion that this find was not a new species of the genus Homo, but a degenerated dwarf form of a completely ordinary modern man who had a genetic defect which had prevented his skull growing to a normal healthy size. Regrettably, during his research, Professor Jacob took the bones to his own laboratory from the safe storage place maintained by the team that had discovered them. He did not allegedly possess the appropriate permits to do so, and when he returned the bones at the end of February 2005 many were slightly damaged and two were even missing. Jacob denied all allegations accusing him of handling an important discovery unprofessionally and claimed that the bones were already damaged when they were entrusted to him for examination. According to him, he handled the remains of the so-called hobbits with the same expertise as any other scientist who had ever dealt with such an important find.

The Indonesian professor's hypothesis that this was not the case of a new species of the Homo genus, but a degenerated variety of members of our own *Homo sapiens* species has been reiterated by several Western experts. So far, however, it seems that most arguments remain on the side of those leaning towards the existence of the new *Homo floresiensis* species. Exact analyses of the preserved hobbit skull enabled them to create a computer-generated model of the dwarf brain structure and reveal that it showed no such characteristics as are typical in people today who had developed a similarly small skull as a result of a genetic defect. Although the hobbit brain was even slightly smaller than that of a today's chimpanzee, it had a significantly different structure. A study of the structure of the skull demonstrated that, in comparison to chimpanzees, the parts of the brain responsible for higher cognitive activities were much more developed in hobbits.

Other discoveries have been made, of which a number of examples exist, that stand against the theory of these hobbits as a pathological form of ordinary humans. Of course, it is difficult to rely on a single 18,000-year-old skull to provide evidence from which the characteristics of an entire population are to be determined, because the being in question could have had an illness which left its bones no longer typical of its species. Fortunately, a number of teeth and parts of jaws have been found, so the doubts of the skeptics can now be dismissed. It would be difficult to explain, for example, how the jaws of two hobbits that lived several thousands of years apart just happened to have the same pathological deformation.

Similarly, some hand bones characteristic of humans are much different in shape than those of apes and older forms of human ancestor. Humans have a very well developed ability to grip things with the thumb and the index finger which is very handy for all kinds of practical activities, but, on the other hand, is not so helpful when it comes to hanging from branches or crushing nuts. It is interesting that the structure of these essential hand bones in hobbits is much more similar to the structure common in apes than to that of humans. This serves as further evidence that hobbits were most likely not a degenerated form of modern humans, but a species of their own.

The final confirmation of the discovery of a new species and not merely deformed humans could only be made possible with an analysis of mitochondrial DNA. However, archeologists have so far been unable to isolate this DNA in the excavated hobbit bones. They hope to do so however with the help of a tooth that has been stored with exceptional care ever since its discovery, so that the remains of the biological material would not be contaminated. If the scientists are able to isolate the DNA in this tooth, we will definitely hear more about it.

The branch that grew away

The discovery of the unusual skeletons on the Indonesian island of Flores is a reminder that evolution is by no means a strictly linear process running directly from simpler to more complex forms of life. This conception of evolution as a vehicle of development is false. Biologists like to tell us that the easiest way to imagine evolutionary development is by drawing an analogy with a tree and its branches. Just as a tree grows into different branches which grow side by side, biological species coming from a common ancestor evolve into different, new species.

The hobbits from the isle of Flores are most likely a species which separated from the evolutionary branch from which modern humans developed around two million years ago, but managed to survive almost up to the time when humans built their first cities.

Before the discoveries on the island of Flores, Neanderthals were believed to be the last species of the Homo genus to have shared the planet with modern man before they became extinct 24,000 years ago. If the age estimates are correct, Indonesian hobbits did not die out until 12,000 years ago, much later than the Neanderthals. Even though experts claim that the chances of that are almost nonexistent, many are excited by the thought that even today, in some remote little island, there could live a group of little people with tiny heads, who can build fires but are not human.

Lucy, more precious than diamonds

It was just another morning in Africa. The paleoanthropologist Donald C. Johanson was sipping his morning coffee with his young colleague Tom Gray, trying to decide where to go fossil hunting for the day. He had been drawn to the remote parts of Ethiopia by his desire to find the skeletal remains of our distant ancestors or at least their close relatives. He studied the evolution of man and other hominids and was now searching for fossils from the period when our forebears had just started to stand on two feet.

His young friend Tom, who was a postdoctoral researcher at the time, had a different mission. With the aid of fossil remains of plants and animals he was to reconstruct as accurately as possible the natural environment which prevailed several million years ago when our ancestors roamed these lands. It is known that Africa used to be much more forested, even in areas where today only desert remains. The characteristics of an environment which existed in a certain area at a certain period could most easily be reconstructed from fossil finds.

Fossil hunting

That day, Donald did not intend to leave the scientists' base camp as he had to catch up on a lot of unfinished administrative work. The day before, Richard and Mary Leakey, two famous and successful human fossil hunters who lived in nearby Kenya, had left the camp after a short visit. During their stay Donald did not spend much time on cataloguing the specimens he had discovered, so he decided to catch up on this work. But, as he later remembered, something drove him to the field that day, even though he was aware that it would be wiser to stay in the tent and arrange his papers. Even so, he wrote in his journal: "November 24th 1974. I'm spending this morning on location 162 with Gray. Feeling good."

Looking for fossils is time-consuming work requiring a fair amount of luck. Many renowned paleoanthropologists have spent much of their lives in entirely fruitless searches: it takes much patience, and a stroke of good fortune, to make a career. This was Donald's third expedition to the Ethiopian region of Hadar and he had already been quite lucky with some of his finds, but he certainly never suspected, that November day in 1974, that he would come upon the trail leading to one of the most important paleoanthropological discoveries of all time.

After they had had their coffee, Donald and Gray got into one of the team's four Land Rovers and drove to site No. 162. Their destination was actually only a few kilometers away from base camp, but they needed half an hour to reach it because of the difficult terrain. When they arrived, the African sun was already scorching hot and all the morning freshness had gone from the air.

The Hadar region in Ethiopia is well known for the riches of the fossils all across its landscape. Hadar is the center of the Afar Desert where there once was a lake in which all different kinds of sediments accumulated. When the lake dried out they surfaced, bearing witness to events that took place in this area several million years ago.

When searching for fossils, the most important thing to have is an eye for distinguishing a potential fossil from a common rock. Then, only a careful and systematic analysis of the area remains to be performed until someone gets lucky. While searching through site No. 162 that morning, all that Tom and Donald found were a couple of teeth belonging to an extinct species of horse, a part of an extinct pig's skull, and a piece of a monkey's jaw. Noon was approaching and the temperatures had risen to more than 40 degrees Celsius.

A three-million-year-old skeleton

After an entire morning of searching for fossils, Tom and Donald decided to head back to their vehicle and drive off for lunch. When they were already on their way, Donald decided to scan the area again, even though others had already examined it at least twice without finding anything. His eyes stopped on a small bone penetrating the surface of the ground. Donald suddenly cried: "It's a part of a hominid's arm!" But Tom did not believe him: "It can't be. It's too small. It probably belongs to a monkey."

However, when they came across a piece of a skull lying nearby even Tom was completely convinced that it was not a monkey, but a hominid. After a more thorough examination of the surroundings, they also discovered some vertebrae and a part of a pelvis. They immediately realized that this was almost certainly one of the greatest discoveries in the history of paleoanthropology. Never before had a scientist succeeded in finding so many preserved parts belonging to an individual hominid of this age. When they found the remains of ribs, they started jumping for joy, in spite of the temperature which was now well over forty degrees.

They carefully marked the site, gathered some more remains of the jaw and set off for the camp to get help. On the way back they picked up two geologist colleagues who were carrying samples of rocks. Tom was so excited that he started to sound the horn before they even reached the camp and started shouting that they had discovered something truly important.

Soon, all of the members of the expedition were on site 162 by the remains of the skeleton that had been discovered by Donald and Tom only a couple of hours earlier. During the three weeks that followed they diligently dug through every part of the area looking for any further pieces s of the skeleton, down to the tiniest shard, which might have survived. In the end, they gathered several hundred fragments of bones which represented approximately forty percent of a skeleton belonging to a single individual.

But who and what was this being they had excavated? At first, the only thing they could be certain was that nobody had ever discovered anything like it. The first night after the discovery had been made nobody went to bed, because the excitement was too overwhelming. They discussed the find into the late hours of the night, drinking beer and going over all the possible implications of their great find. A portable radio played a tape on which the song Lucy in the Sky with Diamonds was repeated several times. Sometime during the night the fossilized skeleton thus acquired the name Lucy and we have known it as such ever since, even though the find's official name is AL 288-1.

Such a small brain, and already walking upright

Lucy was only a meter high and weighed a bit less than thirty kilos. She was classified as a new species of hominids for which the technical term on the site was *Australopithecus Afarensis.* When the world found out about this important discovery made by the researchers in Ethiopia, Lucy became a true sensation. From the form of her pelvis the scientists concluded and affirmed that she had walked on two feet. At 3.2 million years of age she was the oldest preserved specimen of a human ancestor who walked upright.

Only four years later, in a place called Leatoli in the nearby Tanzania, another group of scientists led by Mark Leakey came across a series of footprints preserved in volcanic ash which, mixed with rain, had solidified into a substance similar to cement. When a nearby volcano erupted 3.7 million years ago a being similar to Lucy was one of the creatures who had walked across the wet ashes.

Before Lucy was discovered, scientists were convinced that our ancestors decided to walk on two feet because they had gradually become more intelligent and realized the benefits of freeing their hands for other purposes. With their hands they could do useful things while walking, unlike monkeys. Lucy refuted this theory as she obviously walked on two feet, yet her head was not much bigger than that of a chimpanzee. All the evidence implies that our ancestors had begun to walk upright long before they developed a brain big enough to allow them to figure out that doing so might be a good idea.

The latest research comparing the consumption of energy used for walking in chimpanzees and humans shows that humans with their upright walk use four times less energy than chimpanzees. This means that it was definitely an important development which, among other things, enabled the brain to start consuming more energy. As we all know, people use great amounts of energy even when we sit still and do nothing but think. About one fifth of all the energy produced in our bodies is spent to fuel the brain, which is a lot in comparison to what other animal species use.

To supply our brain with energy, people need appropriate food, preferably cooked so we can digest it more easily and absorb more nutritious substances than we would if we consumed it raw. To obtain meat and other food, rich in calories, we need tools, weapons and knowledge for which free hands and a brain of an appropriate size are essential.

The African Lucy is now on a tour of the U.S. As part of the promotion, scientists in cooperation with the Ethiopian authorities decided to display the fossilized remains of this distant ancestor of ours in a touring exhibition called *Lucy's Legacy: The Hidden Treasures of Ethiopia.*

Paracelsus - Martin Luther of medicine

The life of Paracelsus, one of medicine's greatest reformers, ranks among the most peculiar stories in the history of science. Born as Philippus Aureolus Theophrastus Bombastus von Hohenheim in 1493, only a year after Columbus landed in America, he spent his first years in Switzerland and later moved to Villach, Carinthia, with his family. As both a metallurgist and a doctor, his father instilled into him the secrets of medicine and alchemy of that era. Yet young von Hohenheim soon grew tired of his surroundings. At a mere fourteen years of age, he crossed the Alps on his way to Italy, where he probably spent the first few years studying medicine at the University of Ferrara.

A doctor must be a traveller

After finishing his education – we do not know if he earned a formal degree or not – he embarked on a journey as a 'travelling healer'. Gifted both in medicine and street spectacle, his performance skills always dazzled the crowd, even if they were not always so impressed with his actual treatments. A childhood disease – probably mumps – had also made his appearance memorable. He had no facial hair and a curiously high-pitched voice which made him stand out even more. While travelling, he picked up new medical knowledge and realised that academic medicine was far from omniscient: "A doctor must seek out old wives, gypsies, sorcerers, wandering tribes, old robbers and such outlaws and take lessons from them. A doctor must be a traveller. Knowledge is experience."

Paracelsus means 'equal to or greater that Celsus', which goes to show how self-important our Renaissance champion actually was. In the 1st century, Celsus wrote a medical encyclopaedia which was rediscovered in 1426 and was especially popular in the 15th century. This ancient tome of medicine first went to press soon after Gutenberg's Bible. Its influence can thus in part be ascribed to its availability – printed books had a much higher circulation than manuscripts. It also made an important contribution to medical terminology, giving names which later became standard for many diseases, procedures and anatomical traits. The source of words such as abdomen, uterus and anus can be traced back to this very book. But its content was much less groundbreaking than the scholars thought at that time. The name Celsus thus mostly crops up in relation to Paracelsus.

A new approach to healing

Throughout his life, Paracelsus rebelled against authority, following the example of his role model, Martin Luther. He had practised alchemy, which he probably picked up from his father, since his youth. During his experiments, he for instance noticed that nearby chickens fainted when he was preparing ether - an observation which says a lot about the 'laboratory standards' of those days. He was very passionate about traditional healing methods, based mostly on the 'doctrine of signatures'. According to this belief, the appearance of plants is connected to their healing characteristics. If a plant is shaped like a heart, it cures heart problems; the form tells us what ailment it is best suited for. Paracelsus had no doubt in his mind that this was the highest wisdom of Nature and that a healer must only learn how to recognise such telltale traits and features.

One of the most important innovations he introduced into medicine was the notion that a specific disease requires a specific healing method. Previously, it was a common belief that sickness was caused by an imbalance of the bodily substances, which must be restored to equilibrium in order to cure a person. Paracelsus' key discovery – that every chemical has a distinct effect on the body, which we can benefit from in healing – remains the basis of modern medicine. His travels took him to almost every part of Europe: he is even believed to have visited Scandinavia and England. He also spent time in Constantinople, by that time already under Turkish rule. There, he discovered a book on Byzantine alchemy and a recipe for an elixir based on poppy seeds, which he named laudanum and which was later called opium tincture. For years, he kept the ingredients a secret and only occasionally sold a small bottle to a rich buyer who wanted to ease his pain. To confound potential imitators, he often added a gold leaf into the mixture, which gave him a perfect excuse to spice up the price.

A book-burning professor

Unsurprisingly, this life of adventure took its toll on our healer. Approaching thirty, his hair was all but gone and his body, already disfigured by mumps, was even more damaged due to heavy drinking and frequent tavern brawls. In 1527, Paracelsus arrived in Basel, Switzerland, after a long voyage. There he met the famous Renaissance humanist Erasmus Roterodamus, who was so impressed with the man that he immediately asked him to take care of his own health problems. This Paracelsus gladly did. The most learned man of that time supposedly thanked him by saying: "I cannot offer a fee equal to thy art and thy learning." Taking advantage of his numerous connections, Erasmus tried to find him a proper position that would free him of the need to keep travelling for patronage. Paracelsus was appointed a municipal doctor and gained a medical professorship at the University of Basel. Following in the footsteps of his idol, Luther, he nailed a list of his lectures on the wooden doors of the university for all eyes to see. There, he stated that his lectures were to be open for the public – not a custom at the time – and would be given in German rather than Latin.

For his first lecture, he donned the typical alchemist's garb of a heavy leather apron and announced that he would be revealing a great secret of medical science. Then he theatrically opened a container and revealed that there was excrement inside. Distinguished doctors and other professors naturally started heading for the door, but Paracelsus called after them: "If you will not hear the mysteries of putrefactive fermentation, you are unworthy of the name of physicians." And he was right. By conducting alchemical experiments, he found out that the human body is nothing less than a chemical laboratory, while fermentation is one of the most important natural processes. Only a week after this 'scandalous' lecture, Paracelsus provoked another incident. Together with his students, he built a bonfire at the city square and fed it the books of Galen and Avicenna, the greatest authorities in medicine at that time. Here he once more found inspiration in the acts of Luther, who publically set fire to the papal bull which threatened to expel him.

A genius when sober, a genius when drunk

Despite becoming a university professor, Paracelsus continued to lead a life of debauchery. His secretary wrote about their time in Basel: "Day and night he indulged in revelry and food. He could not be found sober an hour or two together ... nevertheless, when he was most drunk and came home to dictate to me, he was so consistent and logical that a sober man could not have improved upon his manuscripts." The university officials were surprisingly lenient, putting up with his drunken antics for almost two years before throwing him out. Not only did Paracelsus scorn academic medicine, he also looked down his nose at other innovators of that time. His only concern was for his own discoveries; everything else was irrelevant. He for instance opposed anatomical studies of the human body, which later proved to be exceedingly important for the development of medicine.

Despite a decadent lifestyle, Paracelsus fortunately found enough time to write down his findings and publish them in 1536. He died aged forty-eight, in a tavern brawl in Salzburg, a few years after the book came out. The history of science will remember him as a fervent supporter of the assertion that a disease is usually the result of a lack or excess of a certain substance or chemical in the body and not the consequence of an imbalance in the four elements, which used to be thought to constitute the human body.

No more bananas?

This popular fruit is threatened with extinction by a deadly fungal disease attacking large plantations in the tropics. Leading producers are trying to create a new species of banana that would be resistant to the infection, but have not yet been successful. Scientists have offered help, using the latest biotechnology to modify the genetic make-up of the banana in a way that would make it immune to the deadly disease but preserve its original form and taste.

Fruit or vegetable?

Most Europeans and Americans view bananas as a fruit, a dessert, something that accompanies a meal, but this attitude is not shared by the rest of the world. Bananas little different to the ones we know are considered an essential part of the menu by nearly half a billion people in developing countries. The local population does not consider bananas to be a fruit because they have to be cooked or fried before being served as a meal. In local cuisine, they are used as we would potatoes or corn.

Bananas are grown in more than 120 countries on almost 10 million hectares of surface, yielding 100 million tons of crops a year. In developing parts of the world, bananas are the fourth most important food item after rice, wheat and corn. They have a high nutritional value and provide an important source of potassium as well as vitamins A, B6 and C. Because they are easy to digest, they are often the first food suitable for feeding babies. Nearly 90% of all bananas are grown on small farms intended for home use and local markets.

Bananas – the perfect food

The inhabitants of East Africa (Uganda, Burundi and Rwanda) hold the world record in banana consumption, eating 250 kg of bananas per person every year. The local term for banana, "matooke," is a synonym for food. Bananas are also used to produce a popular beer which is an important source of vitamin B for the local population. The advantage of bananas over other crops is that they do not need to be planted anew each year. They flourish in different environments and yield crops throughout the year, which means they represent a constant source of food even when other produce is still ripening. As the urbanization of the third world gains pace and people move from the countryside to towns and cities, bananas are becoming an important source of food in city gardens as well, especially because they are so easy to grow.

There are around 1000 different sorts of bananas that can be classified into 50 large groups. Some banana trees can grow to be 15 meters tall, producing fruits up to 15 centimeters long. The bananas known in Europe and North America are of the Cavendish variety and represent merely a tenth of the total amount of banana crops produced in the world. They are mostly grown on large plantations in the tropics before finding their way to the shelves of our supermarkets. An interesting fact is that Americans seem to eat more Cavendish bananas than any other fruit, consuming more than ten kilograms per person each year. Apples are in second place in the US, with a little over seven kilograms per person every year.

Growing and selling Cavendish bananas is a booming business that involves large amounts of money. The expression "banana republic" denotes large banana producing countries where powerful international companies, such as the United Fruit Company once was, have the deciding say in how the state is run. No step is too drastic for such companies, not even the violent overthrow of a government, where the protection of their lucrative industry is concerned. Among the most notable interventions by "banana corporations" in politics was the removal of the Guatemalan president from power in 1954 and the lending of ships to aid the unsuccessful invasion of Cuba's Bay of Pigs.

The "primordial banana" and its billions of clones

The Dessert Cavendish bananas that we are familiar with are also interesting from a biological point of view. In terms of their genetic composition, the majority of the one hundred billion Cavendish bananas eaten each year are identical. Regardless of how they find their way to our supermarket aisles - from Honduras or Thailand, Jamaica or Columbia - they are all genetic copies of a common predecessor, "the primordial banana" that was brought to a Caribbean botanical garden from Southeast Asia at the beginning of the 20th century. Fifty years ago, this plant became the object of systematic multiplication and mass-scale farming over vast plantation areas to produce the common banana best known today in the west.

The very lack of genetic diversity among the bananas of the Cavendish variety grown on vast plantations around the world represents the greatest danger that one of the world's favorite fruits faces. A disease that can destroy a single banana plant is equally capable of successfully attacking any number of identical plants. An infection affecting one plantation could spread and damage others as well.

The apocalypse of the banana

However odd it might sound, the danger of an apocalypse for the banana is something to be treated very seriously. This is especially true since a large-scale banana plague has taken place before, in the middle of the previous century. A hundred years ago, our predecessors munched on bananas of the Gros Michel variety, which supposedly even bigger and sweeter than Cavendish bananas. At the beginning of the 20th century, however, a fungal infection called the Panama disease started spreading among banana plants. It first broke out in Surinam, then spread across the Caribbean for twenty years before finally reaching Honduras, the world's leading banana producer at the time (today the third largest, after Ecuador and Costa Rica). Hoping to escape the disease, banana growers moved their plantations to new, uninfected areas, cutting down large portions of the rainforest in the process.

By 1960, the migration strategy was abandoned, because it entailed insupportable expenses. That is when leading producers decided to replace the widespread Gros Michel variety with the then quite unknown Cavendish species, which was immune to the Panama disease. Planting and establishing this new banana could naturally only be done at great expense, but it was obvious that there was simply no alternative.

In 1992, however, a new strain of the fungal infection was discovered in Asia, this time harmful even to the Cavendish variety. Since then, the new form of the Panama disease has already devastated banana plantations in Indonesia, Malaysia, Australia and Taiwan, and has begun to spread across Southeast Asia. It has not yet reached Africa and Latin America, but experts agree that it will spread there sooner or later.

How to save bananas before extinction?

Over the last few years, much effort has been invested into saving one of our favorite species of banana from vanishing Scientists can go about searching for a solution in two ways. Following the traditional method, they can try to crossbreed different varieties of bananas in order to create a new species of plant resistant to the Panama disease. On the other hand, modern technology and knowledge enable them to create a resistant plant using biotechnology to modify the genetic make-up of existing bananas. Proponents of the biotechnological approach aim to produce genetically modified bananas which would retain the form and taste of those we know today. The only difference would be that the plant on which the fruit matures would be made resistant to the devastating disease through changes to its genetic code.

The trouble with crossbreeding bananas by natural methods is that they produce an extremely small amount of seeds. For every 300 bananas examined only one seed is found, and then carefully planted in a greenhouse. Only a third of these then sprout. The plants then take two years to grow before scientists can tell if the crossbreeding has worked or not. Then there are further practical considerations. In addition to tasting good and looking familiar, the hybrid banana must withstand long periods of transportation. In fact, adequate replacements for the ailing Cavendish banana have already been found, in terms of taste and appearance: but none met the growers' other specifications. They matured too quickly, for example, or their peel was too thin to endure the long journey from the plantations to our markets.

The banana corporations' principal fear is that the scientists will end up creating a banana which simply does not look or taste quite like the one we know best. People might think a banana is just a banana, but they are in fact accustomed to just one species out of an almost uncountable number of varieties, the Cavendish. Innumerable recipes written with the Cavendish banana in mind may well have to be rewritten. Will we be willing to accept a substitute, as sweet as it may be, of different form, size and taste?

The bioethics of conception

Cardinal Alfonso Lopez Trujillo, president of the Vatican's Pontifical Council for the Family - a panel appointed to advise the Pope on family matters – severely condemned all research on embryonic stem cells in an interview for the Catholic weekly *Famiglia Cristiana* in 2006: "Destroying human embryos is equivalent to an abortion. Excommunication will be applied to the women, doctors and researchers who eliminate embryos [and to the] politicians that approve the law." He became the first high-ranking Vatican official to have proposed the Catholic Church's highest remaining sanction for scientists involved in research on embryonic stem cells.

Anatomy research on cadavers

The call for the excommunication of scientists involved in one of the vital fields of modern science bears a significant meaning. It should at once be put into the context of similar ecclesiastical statements made over previous centuries. In 1300, Pope Boniface VIII issued the bull *De Sepulturis* which runs as follows: "Persons cutting up the bodies of the dead, barbarously cooking them in order that the bones are separated from the flesh, [...] are by the very fact excommunicated."

Many centuries ago, dissecting human cadavers was thus considered to be as morally abhorrent as today's research on embryonic stem cells is to the present Roman clergy. Just as reducing the human body to a subject of medical research raised ethical issues in the past, researching embryonic stem cells presents a problem today. In both cases, one might say, it is a matter of violating the dignity of the human being by using it merely as an instrument for scientific progress.

Body snatchers

In the past, scholars performed most of their anatomy research on animals. The authorities only occasionally allowed them to study the cadaver of some executed criminal. Yet without extensive studies of anatomy on actual human bodies, although a highly questionable practice on moral grounds in the past, one could hardly imagine any progress being made in the field of medicine.

Because cadavers appropriate for dissection were so difficult to acquire, and it was impossible to advance the knowledge of human anatomy without them, the scientists of earlier times simply had to manage some way or another. While writing his fundamental book on the anatomy of the human body, which was released in the same year as Copernicus's groundbreaking work on astronomy, Vesalius was always eagerly on the lookout for cadavers. Some of his memories remain gruesomely vivid: "I stored corpses dug out of graves or collected after public executions in my bedroom for weeks."

Looking back through the history of medicine, one reads of numerous guilty verdicts given for the theft of cadavers for purposes of anatomical research. Thankfully, students of medicine no longer need to creep around cemeteries, digging out fresh cadavers, or wait for executions. (For a period of time, a rule applied in medical schools: a practical anatomy class only took place if students managed to acquire their own cadaver.) Today, however, society has reached a consensus on the conditions that allow a human body and its parts to be used for scientific and therapeutic purposes after a consenting person's death.

The Church changes heart in 1869

Could a similar attitude be adopted towards research on embryonic stem cells? As long as influential people and institutions, among which the Roman Catholic hierarchy and the Church are naturally prominent, persist in regarding a single fertilized cell and a completely developed human being as equals, opinions will continue to clash. Nevertheless, a number of Catholic theologians believe that the Church's attitude will soon change.

Today, it is largely forgotten that the Church has only claimed the same rights for a person and a fertilized egg for less than a hundred and fifty years. It was only in 1869 that the Catholic Church took the view that the life of an individual begin with fertilization. That was when Pope Pius IX eradicated the already established distinction between an unformed fetus, still uninhabited by the soul, and a fully formed being possessing all the human dignity that belongs to a person. A similar distinction between different phases of fetal development, as it happens, was drawn many centuries ago, but only because the mindset for it had already been established in Greek philosophy.

An error in the translation of the Bible into Greek

The discussions about the moral status of the human embryo were heavily influenced by the following paragraph from the Second Book of Moses (Exodus 21), dealing with physical injuries:

> And if men strive together, and hurt a woman with child, so that her fruit depart, and yet no harm follow; he shall be surely fined, according as the woman's husband shall lay upon him; and he shall pay as the judges determine. But if any harm follow, then thou shalt give life for life, eye for eye, tooth for tooth, hand for hand, foot for foot, burning for burning, wound for wound, stripe for stripe!

In the translation of this paragraph from Hebrew into Greek a small error was made. Where the original states that only financial compensation is required in the case of an abortion if the mother is left unharmed, the translators decided to interpret the message in a way which implies that the milder penalty of a fine should only be given if the aborted fetus had not yet acquired a "form", that is, a human form.

The criterion for distinguishing between these degrees of o wrongdoing was transposed from the mother (harmed or unharmed) to the fetus (having form or not having form). The distinction between a formed and an unformed fetus fit well with the widely established Greek idea of the nature of the soul developed by Aristotle. According to Aristotle the soul of a living thing is its form. The interpretation thus prevailed that a formed fetus already possesses a soul. A formed fetus was therefore granted the rights due to a person, while a formless one was merely of economic significance.

The formless and the formed embryo

In the thirteenth century, Thomas Aquinas tried developing a synthesis between the traditional Christian perception of the world and the newly discovered Greek knowledge which, in that period, came to Europe through translations of Aristotle's work. According to Thomas, an embryo does not initially possess a human soul, but merely a form which is also typical of other living beings and enables growth and development. It is only when an embryo develops to the point where it resembles a human being that the rational human soul inhabits it, at the same time making it a human being.

Thomas's interpretation was adopted by the Catholic Church and a similar attitude was also assumed by other monotheistic religions. Until 1869, the Catholic ecclesiastical code distinguished between a formless embryo still devoid of human soul, and as such not a human being, and an already formed fetus, possessing all the features of a human being. Most religions have retained this stance until this very day, so that neither Islam nor Judaism oppose research on embryonic stem cells.

For centuries it was believed that the fetus only develops into a human being when it obtains a soul by taking on human form. So what was it that inspired the Catholic Church to change its moral stance a century ago? Surprisingly, new scientific discoveries about the foundations of reproduction were responsible.

How does an embryo become a living being?

In the eighteenth century, an important debate arose among naturalists on the mechanisms involved in the creation of individual living beings. With the help of the microscope, they discovered that the early embryos of different living organisms look alike, only later developing into distinctly different living beings. So, how can embryos, composed of merely a few cells, know what they have to become?

One theory was that all embryos were completely developed from the very beginning. This meant that all their individual characteristics were already established from the moment an embryo came into being. Pregnancy was considered to be merely a period of growth. This concept of understanding the development of individual living beings was taken to extremes by the philosopher Nicolas Malebranche. If individual offspring were already fully formed inside their mother, this logic would imply that they were also such in their grandmother and all the way back to Eve, she being the first in line of all women ever to bear children.

According to this theory, Eve contained all living beings; they only had to await their actual realization. This is also how Malebranche explained the spread of original sin spread to all mankind. When the first couple broke God's commandment in the Garden of Eden by eating the forbidden fruit from the Tree of Knowledge, all mankind was already present within Eve's body.

As decades went by, naturalists developed this so-called "preformation theory" until their focus shifted from the actual visual similarity between a fetus and a fully formed living being to the presence of a plan which dictated the development of a living being. This idea was the predecessor of today's genome concept.

Problems with the precise definition of the beginning

Similar idiosyncratic interpretations, inspired by ideas such as Malebranche's preformation theory of the embryo, can lead to a rigid defense of the hypothesis that a fertilized egg should already be considered a human individual. However, defining the very moment of fertilization presents some difficulties. When exactly is this new individual supposed to come to life? The new genotype within a fertilized egg is only fully established some 24 hours after a sperm has penetrated the egg. The genes of the new genome do not begin to express themselves until a few days later when the embryo is already made up of several cells.

It is perfectly clear that the embryo at this stage has no form which in any way resembles that of a human being. It is merely a clump of cells, completely incapable of independent development. Even later after the point of fertilization the embryo can split in two, creating identical twins. This occurs only once for every few hundred births, but we do not take it to mean that each twin subsequently has half of the egg's original form and thus half a soul. The idea of one human "soul" per egg is therefore too simplistic. A fertilized egg alone cannot be considered an individual; it should rather be viewed as the matrix of tissue from which individual fetuses may develop.

Just as we succeeded in reaching a social agreement concerning our treatment of the dead, and thus the way in which the body and organs of the deceased may be used, we have to come to a consensus in terms of our attitude towards embryonic cells. The scientific community has already set up its protocols while the majority of the general public all across the world has given its support to carefully regulated research on embryonic stem cells. Now is the perfect time for the Catholic Church to reconsider the moral status of embryonic cells and silence the ongoing reproaches against its obstruction of scientific progress.

Brain

Far out in the uncharted backwaters of the unfashionable end of the western spiral arm of the Galaxy lies a small unregarded yellow sun. Orbiting this at a distance of roughly ninety-two million miles is an utterly insignificant little blue green planet whose ape-descended life forms are so amazingly primitive that they still think digital watches are a pretty neat idea.

Douglas Adams (*The Hitchhiker's Guide to the Galaxy*)

How can I tell you're not a robot?

In the virtual world of the Internet, identities often get blurred. Senior citizens easily pass themselves for teenagers, and timid geeks, normally hiding behind their glasses, can take on the roles of seductive young men and women. If you are not careful enough when chatting with strangers, you might soon find out that the charming young person you have shared so many good talks with is, in fact, none other than your annoying neighbor.

Robots simulating people

The development of the Internet quickly revealed another problem concerning identification, one that can almost be considered a philosophical issue. Not only can people virtually take on other identities on the Web than they possess in real life, computers themselves can impersonate people. Today, this most certainly is not a thing of science fiction as these identity swaps occur on a regular basis.

Robots or, to be precise, programs portraying other people online, sign in to all kinds of Web forums where they post advertisements for their masters or creators. The smarter machines even know how to create their own free e-mail accounts which they use to flood other accounts with advertisements every day.

Exterminator wanted

Robots roaming the Internet and impersonating people first became a serious problem around the year 2000. Administrators of larger websites realized they needed to invent an effective method of distinguishing real people from robots as soon as possible. The task seemed simple enough at first, but it proved to be a rather tough nut to crack when programmers took it on . Should websites have employed people to exchange a few words with every single visitor wishing to sign in? This struck the problem-solvers as an inefficient course of action. A more effective solution was needed.

The job description of a person hired to catch Web-surfing robots would not resemble that of officer Rick Deckard (Harrison Ford) in *Blade Runner*, the legendary maverick who chases runaway cybernetic "replicants" who look and act just like people. Yet a comparison is valid up to a point. The movie implies that officer Deckard himself is a replicant, a computer designed to hunt down other computers. Programmers followed the same principle in designing software which could differentiate between "replicant" and human Internet users. They also sought a method of exterminating these troublesome artificial creatures wreaking havoc on the Web.

The Turing test

Soon after the Second World War, Alan Turing, an English mathematician whose work was integral in the early development of computers, contemplated the possibility of computer thought. He quickly found out this was much too abstract a question, so he shifted to a simpler and more tangible line of inquiry. He started to develop a practical method of differentiating between artificial and human intelligence. How could one determine whether a machine can think? One of the most obvious solutions is to talk to the machine. If it is impossible to tell the difference between its responses and those given by a human being, then there simply is no more difference between artificial and human intelligence.

The Turing test, as it was later named, is, simply put, a method of telling a man apart from a machine. In separate rooms, a robotic and a human subject are posed questions they are obliged answer. Their interrogator, meanwhile, has no notion about which test subject is in which room. If an interview ends with the questioner unable to determine which room contains the living person, the other entity has successfully passed the test and, according to the Turing criterion, cannot be set apart from a real human being.

The Turing test thus offers a way of finding out, through a series of questions and answers with two subjects via a keyboard, if we are conversing with a man or a machine. The machine passes the test when the assessor cannot tell whether he is talking to a man or a machine. The problem for programmers was to design a robotic intelligence which could put the test to other robots.

What robots can't do

Disturbing attacks by malicious robots forced the administrators of larger websites to seek help from experts specialized in artificial intelligence. They quickly concluded that the best solution for "exterminating" such robots would be to set a series of brief tasks that real people could solve in a matter of moments, but which would take computers or robots much longer to complete, if they could do so at all.

Setting mathematical questions was not an option, since computers perform even simple calculations much faster than humans Computers do however have much greater difficulty in recognizing images and symbols. Here human users have a clear advantage: for in the course of several million years of evolution we have become extremely adept at visual recognition, even when the form we recognize is distorted from its usual shape or colored differently.

This is not to say that modern computers are hopeless at spotting an image or symbol they have "seen" before. They can in fact identify letters of the alphabet by a rudimentary scan. This is how they are able to convert the copy of a printed text back into a file which can then be edited and changed anew. However, such recognition is only successful when the image of the text is sufficiently clear and legible. If letters are smudged, text-recognition programs are usually lost. They can only see indistinct blotches where people can still read damaged or smudged writing.

The Automated Turing test

This was the disadvantage engineers sought to exploit in creating a robot which could differentiate between artificial and human intelligence. Humans might wince a little when presented with scrappy handwriting; but computers are brought to a standstill by it. Researchers at Carnegie Melon University therefore developed an automated Turing test requiring no human guidance. They named it CAPTCHA: the Completely Automated Public Turing Test to Tell Computers and Humans Apart.

In simple terms, CAPTCHA is a program that can compose and evaluate a test which it could not actually solve for itself. It is a program for exterminating programs. It gives users a simple visual task that usually requires them to spell out a series of smudged letters and numbers shown on the screen. On the basis of the answer given, the program can tell whether visitors to a website are human or electronic.

Every Internet user has probably encountered such a program at some point. We have to pass its test in order to register on an Internet forum, to create a new e-mail account or to file an income tax return. With time, of course, artificial Web-surfers have become much better at reading unclear print, and programs for exterminating programs must be constantly improved. A delicate balance must be struck here: CAPTCHA must not make the letters in its tests so hard to make out that they baffle humans and computers alike.

Which is the best school for computer science?

In November 1999, a good while before the first online robotic robot-catchers were developed, the website slashdot.com conducted a survey to find out which US university offered the best course in computer science. This was obviously quite a delicate question, especially when posted on the Internet.

To make voting as fair and free as possible, each computer was given one vote. The program that registered the votes also registered the Internet addresses (IP addresses) of individual computers that had already cast their vote, allowing each individual to vote only once. However, students from Carnegie Mellon University found a way to bypass this safeguard and designed a program which could cast thousands of votes for their university. By the next day, MIT students had written their own voting program; and so began a battle between the two institutions to generate the greatest possible number of non-existent and fraudulent votes. When the poll closed, MIT snatched victory by 21.156 votes and Carnegie Mellon took second place with 21.032; while all of the other candidates languished far behind with less than 1000 votes each.

Certainly the students of the winning school and the runner-up made a strong case for the quality of their respective faculties. When it comes to deciding on the best school for computer science it might even be thought appropriate to grant computers an equal right to vote; although only, perhaps, if they could pass a Turing Test first. In the case of other surveys, however, one would probably be more comfortable knowing that each registered vote represents a human being.

The man with no memory

When Clive Wearing developed an excruciating headache on March 26th 1985, it would not have been surprising if the cause had been mental overwork: Wearing was a renowned British musicologist, BBC producer and choir director. Having taken some painkillers with little effect, he stayed at home to rest, his doctor having assured him that he had simply caught a bad dose of flu. A few days later the pain had still not subsided. He was unaware then that these would be the last days in which he would be able to remember in the evening what he had been doing in the morning.

When his wife returned from work on the fourth day of his illness, their apartment was a mess and her husband was nowhere to be found. Naturally, she immediately became worried because she remembered he had still had a high fever in the morning. Clive was spotted aimlessly wandering the streets by a cab driver who informed the police. Clive and his wife spent the rest of the day in hospital, where he underwent a series of examinations and was diagnosed with a severe case of a viral brain disease called herpes encephalitis.

They started treating him with antiviral drugs, but the virus had already severely damaged his brain. Because of the infection a great part of his brain had already been destroyed, so that on the image his head appeared as if entire pieces of it were missing. The greatest damage was inflicted on the very part where memories are formed and stored. Even though Clive survived the viral infection and later completely recovered physically as well as mentally, he never regained his memory. Today, more than twenty years after the infection, he is believed to be the man with the worst memory ever to have been thoroughly medically examined.

A Prisoner of the Present

Since 1985 Clive has been "waking up" from a long sleep every few minutes, not knowing where he is. Most of us have probably experienced the unpleasant feeling of waking up and at first not knowing where or even who we were or what was happening. After a few seconds, though, we are usually able to bring back to mind the events of the night before and recall how we ended up in this seemingly unknown bed. Unfortunately, Clive has no memory to help him and finds himself completely lost in time again and again when after a few minutes - at times no more than seven - his short term memory is erased.

Of course his mental capacities have remained almost as strong as they were before the illness occurred. He can still be a very intelligent and witty conversational partner. Unfortunately, everything that happened after 1985 is completely new to him. On being introduced, he would forget who you were after a few minutes unless he knew you from before. Usually he spontaneously understands this absence of memories to be the result of a coma from which he had woken after a very long time. He feels as if he had only been awake for two minutes. Day after day, he poses the same first question to people he finds in his company: "How long have I been unconscious?"

For a while, he used to keep a diary which contained almost nothing but an exact record of the time and the note: "Now I am awake." The fact that he was awake, that he was conscious of his existence, was what he thought to be the most important piece of information that he had to store each time he "woke up" again. Of course, he could never understand exactly why the previous record in his diary which was also written in his handwriting said: "Now, I am really awake." That is because the record also contains the time shown on the clock only some ten minutes earlier. As he can never remember writing the previous records, he is often convinced that they are incorrect, so most of the entries in his notebook are crossed out. Sometimes he only corrects the time given by the entry, because the record stating that he is awake seems irrefutable, while only the time written beside it contradicts the hour on the clock he sees before him.

Clive has been the subject of several documentaries and some clips taken from them are available on the internet. The most surprising thing one notices when watching these excerpts for the first time is the intensity of the feelings Clive expresses for his wife every time he sees her. Even when she returns to the room after only a couple of minutes he reacts as if she had been away for a decade. In fact, his love for Deborah is the same or even stronger than it was when he fell ill, not long after their wedding.

Tales of Music and the Brain

Like his intense love for his wife, his capacity for creating music is another thing from his previous life that has remained unchanged. Soon after his loss of memory, Clive's wife noticed that he could sing or play pieces of music exceeding the span of his conversational memory. It seemed that, to him, singing was even easier than speaking. When playing the piano he is completely taken over by the music. This lasts until the end of the song when the brain mechanism enabling him to play music switches back to the mechanism that controls the processes of thinking and linguistic expression. Between these two phases healthy people would usually search through their memories and remember where they were and who they were playing the song to. Unfortunately, Clive does not possess this ability and, when the song ends, he finds himself utterly confused, not knowing where he is or how he got there.

In a rehearsal recorded with the choir he led before he fell ill, he can be seen completely focused on conducting the choir and instructing the singers when to join in. Once the song ends, though, he suddenly starts violently shaking and is completely confused. This is when his memory should have come to his help, but he is simply unable to activate it.

The discovery that Clive has retained his talent for performing music despite his severely injured brain is a matter of ever increasing interest to scientists researching brain activity. The renowned neurologist Oliver Sacks even gives a detailed account of Clive's story in his book, *Musicophilia: Tales of Music and the Brain*. Sacks made his name as a psychiatrist in the seventies when he succeeded in waking up some of his patients from a state of numbness, an achievement which he describes thoroughly in one of his widely acclaimed books. The famous movie *Awakenings* (1990), in which Sacks was portrayed by Robin Williams, was based on this book. In addition Sacks has written about his unusual patients in many other works which are also masterpieces of narrative.

His style of writing is probably best represented by a passage from his collection, *The Man who Mistook his Wife for a Hat*, published in 1985:

> There was a hint of a smile on his face. He also appeared to have decided that the examination was over, and started to look round for his hat. He reached out his hand, and took hold of his wife's head, tried to lift it off, to put it on. He had apparently mistaken his wife for a hat! His wife looked as if she was used to such things.

Sacks' writings are invariably about people who could not tell their own stories, unless in moments of exceptional clarity; and this is a principle which is all the more marked in the case of Clive Wearing.

Newly in Love Every Day

Clive spent most of the six years after the virus had destroyed his memory in a psychiatric hospital. He became very depressed as, even though he was, in a way, completely healthy, he was unable to lead an independent life. Day after day he woke up in a strange hospital room with unfamiliar doctors and hospital staff. Nearly seven years later and with the help of his wife's persistent persuasion, he was transferred to a smaller countryside clinic for patients suffering from brain damage, and there he immediately started to feel much better.

In 2005 his wife Deborah wrote a book about her experience with her husband's illness, entitled *Forever Today: A True Story of Lost Memory and Never-Ending Love* in which she described the good as well as the unpleasant moments of their relationship. She was 27 when Clive fell ill and although the first years of his condition were extremely difficult for her, she tried to do her best with a situation which offered no hope of improvement. In 1992 at 35 years of age, however, she came to realize that she could not go on living like that anymore. She decided to divorce her husband and move from London to New York. She wanted to start a new life. Of course, Clive knew nothing of the divorce and even if he did, he would have forgotten it only a few minutes later. The divorce papers were signed by Clive's son from his first marriage who is also his father's legal representative.

Naturally, Deborah still visits Clive, who continues to be immensely happy every time he sees her. As he has no sense of time, his feelings for her are as strong when he sees her after a few months as when he sees her coming from the bathroom after only a couple of minutes.

She's blind, but she sees

Defective gas water-heaters can be extremely dangerous. An Englishwoman called Dee Fletcher had the misfortune to find this out herself in 1988, while taking a shower in the house she and her husband Carlo had recently bought in a small village north of Milan. Unaware of any possible hazards she decided to freshen up in her new bathroom and carelessly started to shower, without the faintest idea that deadly gasses might be filling the air. The room was poorly ventilated, giving the water-heater too low a supply of oxygen to burn; instead, it began pouring out a cloud of deadly carbon monoxide.

This poisonous gas is all the more lethal because it is not only invisible but odorless, so someone inhaling it usually notices nothing until it is already too late. Unfortunately, this is exactly what happened to Dee. She lost consciousness and would have died soon afterwards if Carlo had not returned home in the nick of time. He immediately started to resuscitate her and quickly got her to the nearest hospital, where they saved her life. At first, nobody knew how long her brain had been deprived of oxygen, and thus whether permanent damage had been caused. The brain is the organ which suffers the most rapid deterioration without oxygen. Brain cells start dying if the air supply is broken even for just a few minutes.

She can't distinguish objects, but she sees their details

When Dee woke up in hospital the doctors soon realized that the carbon monoxide poisoning had left its mark on her brain. Although she could speak normally and understand what others were saying, she could no longer see. The doctors' first diagnosis was that the poisoning had destroyed the sight center in the brain, but they had to change their opinion in the next couple of days as Dee's vision seemed to be returning. When Carlo came to visit her she mentioned that he was wearing the same sweater as the previous day. She also knew that outside the sky was clear, and that the flowers in her room were red and blue.

Yet when her mother arrived from England Dee could not even recognize her when she first came into her hospital room. It was only on hearing the stranger's voice that she knew her mother was there. On the other hand, Dee had no difficulty handling objects and recognizing colors. The next day, when she and her mother were having a cup of coffee, she noticed something more unusual still. She was startled by the sudden extreme clarity of her perceptions: "You know what's strange, mom? I can see the little hairs on your arms perfectly clearly!"

Her mother was beside herself with joy when she heard her say this, because she thought that her daughter would gradually regain her sight after all. Yet the happy moment was short-lived. It became evident that Dee could no longer recognize the shape of the objects she was looking at. Even though she could discern the color and texture of a thing, she could not tell what its nature or function might be. The poisoning had caused Dee to lose only a part of her ability to see. While she could still perceive the colors and the structures on the surface of individual objects, she could no longer make out their shapes.

Blind to shapes

Her blindness was later studied by English neurologists who discovered that she was still able to distinguish colors very well. She was capable of recognizing very small nuances in the shades of individual colors. She could also tell if an object was made of plastic, wood or metal, but she could still not discern shapes. When she was shown parallel lines on a computer screen she had no trouble recognizing the pattern she saw, but she could not determine whether the lines were vertical or horizontal.

It is necessary to state that she had no previous problems with her eyesight which, of course, could have caused her to have blurred vision like someone who is very short-sighted. Her vision was completely clear and sharp which is supported by the fact that she could recognize from a distance the colors and patterns that myopic people could not see. After thorough examination it became clear that Dee was only blind to shapes, but otherwise had perfect vision. It does not matter what the shape in question might be, Dee is unable to recognize it. The problem she has is not with comprehending what she sees, but with the process of looking itself. It is not a matter of her brain recognizing shapes, then failing to interpret them and link them to words. Dee simply cannot see shapes.

When doctors asked her to draw an apple from memory she did very well, but when they gave her an image of an apple to copy, she could only sketch out a few indistinct scribbles. During the first years following the accident she also had nightmares which were quite different to the usual variety of bad dreams. For her, the nightmare began when she woke up. When she dreamed, her sight was fine, but when she woke up she returned to her personal shapeless world.

The medical term for the condition in which people are no longer capable of recognizing the objects about them, even though their eyes function normally, is called "agnosia". The term was coined at the end of the nineteenth century by a then still little-known neurologist named Sigmund Freud, who used it to describe patients who had difficulties making sense of what they saw. Of course, there are many different kinds of agnosia, depending on which part of the patient's brain is damaged. In medical literature, many types of this "conceptual blindness" have been described, some of which are also quite exotic, such as amusia, which deprives sufferers of the ability to recognize music, even when their hearing is quite unimpaired.

Dee's condition is a case of the visual form of agnosia. However, it is not a classic example of the disease. Neurologists were surprised to discover that Dee was completely capable of picking up a pencil, though she was not able to recognize its form. Regardless of how the pencil was placed on the table or which direction it faced she could pick it up effortlessly. This implied that she did in some way perceive shapes, but was not aware of it. If she were completely blind to shapes she would not have been able to pick up the pencil so easily.

Do people have two separate brain mechanisms for seeing?

When researchers delved deeper into Dee's case, they became increasingly aware that Dee could still respond very well to her environment, as if her eyesight was unimpaired, but at the same time could not make sense of the forms she was actually seeing. What did this mean? Researchers made the hypothesis that people do not have only one, universal mechanism that gives meaning to what we see or, in other words, the visual data that travels from our eyes to our brain via our nerves. Dee's case of agnosia clearly suggests that in our brains two separate processes of visual information analysis are at work: the first is responsible for coordinating movement and is independent of the second, which creates an inner image of the outside world. In Dee's case the poisoning accident damaged the second mechanism, while the first continued to function normally.

The theory of two modules for processing visual perception in the brain has been developed by two British neurologists, Melvyn A. Goodale and A. David Milner, and presented in their book *Sight Unseen: An Exploration of Conscious and Unconscious Vision*

(Oxford University Press, 2004). In it they give a detailed description of the case of Dee Fletcher who they examined and tested on several occasions following her accident. In the book, they also discuss cases of patients with the inverse combination of symptoms of those Dee has. In some cases, meanwhile, patients may recognize shapes but not colors. However, this is not a classic example of color blindness where the color receptors in the eye are damaged. These patients are capable of effortlessly passing the classic color blindness tests, because they can distinguish the borders between two different colors of the same intensity, which the typical monochromats cannot, but they are still unable to tell which color they are seeing. When asked to draw a banana, for example, nothing would stop them from choosing a red or a green crayon instead of a yellow one.

Fifteen years after the accident the two scientists went to visit Dee and her husband Carlo at their home. They say that when Dee came to open the door, she showed almost no signs of blindness at all. She has no difficulties moving quickly her house, cooking and even working on the garden, as she was also glad to show them. When they went for a short trip the following day she seemed to follow the forest path easily, and only needed occasional help in keeping to the right trail.

Although she never regained perfect vision, Dee coped by optimizing her use of the sight mechanism which remained unharmed. She now tells people apart by the color of their clothes, skin and hair. She uses the same technique when doing household chores which she now performs considerably better than immediately after the accident.

How babies learn languages

It is well known that a person who has been raised in Japan and whose mother tongue is Japanese cannot tell the sound *r* from the sound *l*. If such a person were to hear the sentences *He likes to read* and *He likes to lead* taken out of context, he would most likely not be able to tell them apart. To him, both sentences would sound the same, because in his early youth his brain became accustomed to the Japanese language, in which there is no difference between the two sounds. Similar peculiarities in perceiving the differences between sounds also occur in other language environments. The Spanish and the French distinguish the sounds *b* and *p* in a different way than the English. What sounds as a *b* to a Spanish person, would sound as a *p* to an English person. Meanwhile a Thai person would hear three sounds in *b* and *p* where English-speakers hear only two.

The basic units of verbal sound are called "phonemes". Every language has a set of such units, but none can include an exhaustive range of phonemes, covering the full scope of possible sounds the human mouth can make. The structure of the brain which stores the phonemes any person is able to hear becomes fixed very early in life. Most people can therefore only make out the basic sounds they have heard in their first linguistic environment.

Suckling on consonants

We all know that little children immediately pick up the language of the environment in which they are being raised, while later one has to try much harder to learn a new and unfamiliar language. Why is there such a difference? Would it be possible to activate the ability of spontaneous language acquisition later in life? Questions about how the brain acquires and understands and then uses a language have always fascinated scientists in this field of research.

Because it is impossible to ask babies whether they can tell the difference between, for example, *read* and *lead*, scientists have come up with a great many alternative methods to test whether infants can differentiate between individual sounds, and find out when they start to lose this ability. They created, for instance, an artificial breast which did not produce milk when a baby suckled on it, but instead emitted sounds. The more the babies sucked this speaking breast the more distinct sounds it produced, and the children were almost as pleased by this as if they were actually being fed.

However, if the children could only draw a single repeated sound out of the breast, they would eventually tire of sucking on it. When the sound changed, though, they would start sucking more actively again. It was this change in the intensity of sucking which revealed that children could tell the difference between two sounds, otherwise such a reaction would not have occurred at all. This made it possible to observe whether little children could differentiate between the sounds *r* and *l*, despite the fact that they were being raised in the Japanese language environment.

At first, scientists assumed that the children would not be able to detect the subtler differences between individual sounds, but that they would gradually learn to do so as they grew up. Surprisingly, they discovered that in fact the truth was just the opposite. Even babies of no more than a month old babies from an English language environment were able to distinguish all the phonemes that were characteristic of the English language. These babies had already developed an English-language framework within which they could tell the difference between all the different *r* sounds and all the different *l* sounds. They could not however detect the differences *within* both individual groups of sounds, even though they were pronounced by different speakers.

The surprises kept on coming. When the research team tested babies from other language environments, they found out, for example, that Mexican children could easily distinguish between the different sounds of the English language. Similarly, Japanese babies had no difficulties with telling apart the sound *r* from the sound *l*, despite the fact that their parents were unable to do so.

Babies can speak all languages

Infants can distinguish between phonemes of all the existing languages of the world even when they have never heard a given language before. It makes no difference whether the sounds are characteristic of French, English, Chinese or Slovenian, infants have no difficulties telling them apart regardless of who pronounces them, be it a man, a woman or a child. Babies have an inherent aptitude for learning languages. In fact, they have a sort of universal talent for learning any of the approximately 6000 languages that still exist today.

It is also important to emphasize that a child does not merely distinguish individual sounds like a computer detector which can be set to detect the difference in the frequency of two consecutive sounds. Children are able to correctly categorize sounds by classifying them into groups on the basis of which words are formed. When born, every child functions like a sort of universal receiver that can acquire any language. The key question that arises is why babies, which are born as universal linguists, later become specialized in their language only, and lose the ability to differentiate between the phonemes of other language environments.

Because the breast imitation test was not so effective with older infants, scientists developed other methods to monitor how babies respond to different sounds when they pass six months of age. Today, electric sensors are used for this purpose, attached to the child's head in a sort of hat to measure minute differences in brain activity. These devices can detect when the child notices or fails to notice a difference between two phonemes.

In Japanese children, researchers discovered that they could still easily distinguish between the English *r* and *l* at seven months, but would lose this ability only three months later. Nine-month-old Japanese children could no longer tell the difference between the two sounds while American children of the same age actually became more observant of this difference. Similar research conducted on Canadian children revealed that at six months of age they could hear distinctions between the phonemes of exotic languages which were inaudible to their parents and twelve-month-old peers. By the age of one, however, children lose this ability to learn perfectly any language they hear. Instead their brains adapt to deepen and broaden their ability in the language of their everyday environment.

Scientists have observed a similar adaptation in deaf children. Deaf infants go from being able to acquire the elements of any sign language whatsoever to developing a more complex facility with a single system. A number of research paths thus led to the conclusion that all children at about the same age become much more specialized in distinguishing basic parts of speech. Research has shown that once they have lost their ability to differentiate between all phonemes, children begin to learn the language of their environment more rapidly than those who remain sensitive to the sounds of languages for a longer period of time. The reason for this is that communication is substantially easier when the brain focuses on the differences important for conveying information and ignores everything else.

The time when children start losing the ability to tell the sounds of foreign languages apart approximately coincides with the time when they redirect their attention from sounds to words. That is when they first become familiar with all the possible words in their language environment. Researchers have found that nine-month-old children prefer listening to sets of sounds which combine in the same ways that sounds usually combine in their native language, even if those sounds do not form actual meaningful words in that language.

Direct contact is important

The latest research also puts fascinating emphasis on a child's need for direct personal interaction in learning a language. The importance of personal contact emerged in an experiment to find out whether exposing children to a foreign language could prolong their early ability to differentiate between the full range of phonemes.

A number of nine-month-old infants were separated into four groups. All of the children had been raised in a purely English-speaking environment. Over the course of a couple of weeks, the children in the first group played with Chinese teachers who communicated with them solely in Chinese. The second group watched the Chinese teachers on video, while the third group only listened to them and the fourth group had no contact with them whatsoever.

The experiment showed that only the children who had been in direct contact with the teachers preserved their ability to tell Chinese phonemes apart later on. Their peers in the control groups who had only listened to the teachers or watched them on video could later distinguish the foreign phonemes no more than the children who had neither seen nor heard the Chinese at all.

Why people are exceptional readers

Reading and writing relatively recent inventions, no more than some thousands of years' old, and are much younger still in some parts of the world. People can read and write with ease and it seems almost by instinct, using the same eyes and brains as our ancient illiterate ancestors. Reading is not, however, an evolutionary feature of the human brain: if it was, then prehistoric Homo sapiens would also have been able to read. The fact that modern people are such good readers thus presents something of a mystery.

When reading, we are not actually aware of individual letters, because written words and sentences are automatically transformed into mental images. When we read a well written novel, we are completely immersed in the story and events take place before our eyes as if watching a movie. Of course, another important factor that makes a good book is the flow of words which create the atmosphere. However, it is essential that this is achieved spontaneously, without any forced conscious effort. As readers, we often do not realize that our eyes are merely registering a sequence of signs which is then simultaneously translated into meaning by our brain. It seems as if the writer were whispering in our ear.

Humans started communicating gradually via speech, through evolution, which means we are biologically conditioned to speak, but this does not apply to reading and writing. In his latest book, *The Vision Revolution: How the Latest Research Overturns Everything We Thought We Knew about Human Vision* (Benbella Books, 2009), Mark Changizi, an American scientist, discusses the paradox of reading in the history of human adaptation.

Children's drawings, the origins of writing

In the middle of the 20th century, child psychology researcher Rhoda Kellogg discovered that little children of the same age and from different cultures and parts of the world made similar drawings. The goal of human evolution was not to draw well, so according to Changizi what these findings imply above all is that children have a universal approach to learning to communicate to the best of their abilities, using a pen and a sheet of paper. When drawing, children are creating signs they believe to be visual images symbolizing individual objects in their surroundings.

The key word to bear in mind here is "symbolizing": the child in his or her drawing produces an image which "stands for" something they have encountered (or imagined), not one which aims at describing or representing that entity perfectly. As one might expect, children's drawings from different parts of the world are not completely alike, even when they are pictures of such common favorites as cats or dogs. An element of style is already important: the same variant, in fact, which one observes between art from different cultures.

The key feature which all children's drawings have in common is that they are above all symbols and not attempts at actual reproductions of real-life subjects. As such, children's drawings represent an effort to write down stories using graphic symbols. They are stylized versions of real subjects and are, much like cartoons (which are a sort of upgrade on children's drawings), suited to children's needs to portray a story than to give exact images of reality.

However, symbolic communication as developed by children in their drawings is not very effective. The trouble with this type of "writing" is that it is not comprehensive enough to encompass the diversity of all the words in our vocabularies. Effective communication would require several tens of thousands of different symbols – a practical impossibility. Another problem is that our visual system analyzes images in several successive phases. The brain first analyzes simple features or graphic elements, after which it focuses on the relations between these elements. The objects that we ultimately perceive are only constructed from these partial analyses in the last phase.

The mechanism of object identification

Humans have an innately well-developed mechanism for object identification. The brain has developed the ability to recognize different elements of the outside world very effectively. Our brains always try interpreting any image we perceive as being one we are already familiar with. We thus always try to see in terms of things we recognize and comprehend. The brain's initial analysis of the visual elements before us takes place without us being aware of it: we only "perceive" things the brain has rendered intelligible as completely identified objects.

According to Changizi, writing systems work on the basis of this natural mechanism for identifying forms. The letters of different writing systems were supposedly created to correspond as much as possible to the basic graphic elements by which our brains identify objects and their positions. To the brain, letters are like the constituent parts of objects which it builds up and interprets to form a comprehensible image. Words, meanwhile, are analogous to the objects formed from such smaller units. The overall process is facilitated by the brain's capacity to identify and interpret images it recognizes. In looking at words, this capacity to recognize is put to a new purpose: instead of perceiving an object it identifies as similar to something it has seen before, we relate a graphic symbol to a meaning or value with which it has already been connected. In much the same way that our brains analyze the parts of objects and then identify the objects themselves, we are unaware of perceiving individual letters and forming words from them, but seem to pass directly to the meanings of the words.

The forms of letters are not random

If words resemble objects, as they do in simple writing systems, then their smaller constituents, letters (or "graphs") must resemble the visual elements which the brain identifies in those objects. This principle has practical complications. Firstly it sets a strong limit upon the values or meanings a writing system can s express – for how do you write a sign which "resembles" an abstract concept? Secondly, it places severe demands on the skill of the average writer or reader. For there are many objects which are very complex to represent and thus difficult to recognize in graphic form: yet to be successful, writing must be easily and speedily processed by the part of the brain which interprets and recognizes visual forms.

The answer to these problems was found through the route adopted by languages in most parts of the world. Today, most writing systems are based upon the written word. This enables us to use a substantially lower number of symbols, because languages usually only have several tens of different phonemes which roughly correspond to the letters of the alphabets they use. These graphic symbols retain only the most rudimentary resemblance to the basic visual elements the brain identifies in objects we see and recognize. As with children's drawings, this resemblance is in any case highly stylized. There is a relationship nevertheless between the frequency with which certain letters are used and how commonly their corresponding visual units are identified by the brain in the world around us. The outlines of objects and their intersections usually have the form of the letters T, L and K. But the form X, for example, occurs much more rarely, so the letter X is less frequent in writing systems.

Changizi analyzed all different kinds of writing systems he could come across in the history of different cultures around the world. He realized that some of the fundamental structures of all writing systems were alike and, as such, universal. He found that the forms more commonly perceived in nature were also better represented among human graphic symbols.

He also reflected on how the brain's incredible adeptness at analyzing visual signs could be used for other purposes. We recognize the objects in our surroundings instinctively and without any effort, while other tasks are much more demanding. Wouldn't it be wonderful, then, if a difficult mathematical problem could be transformed into an image, so that only a glance would suffice to reveal the solution? Diverse modern applications have already been found for the brain's versatility at recognizing graphic symbols.

The CAPTCHA scheme for identity verification on the internet tests and exploits precisely this ability when it asks website users to transcribe a word put before them in smudged or distorted characters. No artificial intelligence has yet been able to match an ordinary person's powers of visual recognition. It could thus be said that this faculty which allows us to identify objects and decipher script is a defining feature of the human mind.

Watching the birth of a new language

In 1769 the Berlin Academy of Sciences offered a prize for the best response to the following question about the origin of language:

> Supposing that human beings were left to their own natural faculties, would they be in a condition to invent a language? And, if so, by what means would they manage to invent it? We are looking for a hypothesis that would explain the problem in a clear manner and give an answer to all its different aspects.

The winner of the prize was Gottfried Herder with his *Essay on the Origin of Language*. Herder reveals his approach in the very first sentence of his essay: "Already as an animal, the human being has language." It is unlikely however that any of Herder's fellow contestants would have expected that, one day, people would observe with their own eyes the spontaneous birth of a new language, an event which took place just a few decades ago.

The experiment of king Psammetichus

For several centuries, the question about the origin of language was one of the central subjects for debate between philosophers and other scholars. The interest in the mystery of the nature of language reached its peak in the Age of Enlightenment when the great questions about nature and the origin of man came to the forefront of scientific thought. Is language innate or acquired? Why are there so many different languages in the world if language is a gift of nature? What kind of language would children develop if they were left alone on a desert island after being born?

The question about the origin of language can first be encountered in the story about the unusual experiment carried out by the Egyptian king Psammetichus, the account of which may be found in Herodotus' *History*. Psammetichus wanted to find out whether

Egyptians truly were the oldest people on the face of the Earth, but was "despite his efforts unable to find a way which would put him on the trail of the original man". In order to solve the mystery, he conducted a true scientific experiment:

> He took two newborn babies from randomly chosen parents and entrusted them into the care of a shepherd who had been instructed to keep them in his flock. The shepherd was ordered not to speak a single word in their presence and to leave them lying alone in an isolated space, gradually accustoming them to two goats. After the children had had their share of milk the shepherd would go on with his chores.

From Egypt to Nicaragua

King Psammetichus isolated the two children from all cultural influences

> because he wanted to hear what word the children would utter first after progressing from inarticulate babbling. This is also what actually happened. Two years after the beginning of the experiment when, one day, the shepherd opened the door and entered the room, the boys threw themselves at him, hands reached out, calling out *Bekos!* /.../ when Psammetichus heard them with his own ears, he wished to know which language the word *bekos* belonged to and what it meant. He found out that it was the Phrygian word for bread. From this experience, the Egyptians concluded and admitted that Phrygians were older then them.

Even though many scholars over subsequent centuries probably considered carrying out an experiment similar to King Psammetichus', this could naturally not be allowed. That is why scientists were even more excited a few decades ago when an event comparable to King Psammetichus' experiment took place quite spontaneously. In addition, the children that happened to be involved in this unusual "experiment" even gained from it, and none of them suffered like the boys in Herodotus. This time the story took place not in Egypt, but in Nicaragua.

Opening a new school for deaf children

At the end of the 1970s, the life of deaf children in Nicaragua changed immensely. When the Sandinista came to power, one of the changes they introduced was the reform of public education throughout the country. In Managua, they opened a new school center for educating children with special needs. Before, parents usually kept deaf children at home because, at the time, deafness was stigmatized. Within families, communication with deaf children usually consisted of a few simple gestures which were no more than the bare essentials required on a daily basis.

In the capital of Nicaragua, a smaller school trying to teach the deaf to speak Spanish had been open for several decades, but with no great success. It was only when the larger school was established that, for the first time, deaf children from different parts of the country were brought together in substantial numbers. Gathering this large group of children who had no effective means of communication proved to be the key element leading to the creation of a new language.

The children can talk, but the teachers can't understand

The teachers tried teaching the children using a Russian method based on a visual alphabet system recommended by Soviet instructors, but this was quite unsuccessful. However, the children started spending time together outside the confines of the school, and exchanged visits like other ordinary children. To their surprise, the teachers gradually started to realize that the children were very good at communicating with each other. Yet while they could communicate with one another very well using gestures, the teachers had a difficult time understanding them.

In June 1986, the Nicaraguan Ministry of Education contacted an American scientist, Judy Kegl, an expert on the linguistics of sign language. She was invited to visit the school for the deaf in Managua and to try and find out how these children were communicating. Needless to say, the 33-year-old professor immediately embarked for Nicaragua.

Witnessing the "Big Bang of a language"

Kegl quickly discovered that the children were not using any of the many established sign languages, but had developed an entirely new way to communicate using gestures. After a few days of observation, she managed to learn the sign for "house" or "home" and to ask the children where they lived. Each of the children responded with a series of complex, at first sight completely meaningless hand gestures. It was not until later that she discovered that the series of gestures actually described in great detail how to take the bus to each of the children's homes.

At the beginning, she faced many difficulties because she could not make out the grammar of the language used by the deaf children. Later she remembered how, at first, everything seemed very chaotic. After three weeks, she also visited the school for younger children. On her first day, she watched a girl in the school courtyard accompanying her signing with quick and rhythmical hand gestures. Kegl realized the child was not simply mimicking the signs she had observed in older children: "I looked at her, and I thought to myself, Holy cow, that girl is using some kind of rule book." Ann Senghas, at the time Kegl's assistant, remembers this event with great excitement: "It was a linguist's dream. It was like being present at the Big Bang."

What do we know about the nature of language?

Using a special universal system for writing down gestures also known as sign writing, which is used by about three thousand people, several books and textbooks have already been translated into the language of the Nicaraguan deaf children. In addition, much research has been carried out and many scientific discussions have been written, tackling essentially the same key question the Berlin Academy of Sciences had set for its prize in 1769. To this day, scientists have yet to reach an agreement as to what extent language is innate or acquired.

Either way, the discovery of the spontaneous birth of a new deaf language in Nicaragua has opened the way to many new findings of great help in our search for the answer to the great question about the nature of language, a question that humankind has been contemplating ever since the times of the Egyptian king Psammetichus, if not even longer.

Pseudo-patients in psychiatric hospitals

In January 1973, the prestigious journal *Science* published a report on an unusual experiment which the American psychologist David L. Rosenhan of Stanford University had carried out with his assistants. He wanted to know how the staff of psychiatric hospitals treated their patients and how effective the doctors were in making psychiatric diagnoses. The article entitled "On Being Sane in Insane Places", in which he gave a detailed description of the results of his experiment, is believed to be one of the most resounding articles in the field of psychology because it initiated along-lasting polemic about the validity of psychiatric diagnostics.

Pretending to hear voices

Rosenhan, four male and three female colleagues, of which three were psychologists and the others were a pediatrician, a psychiatrist, a painter and a housewife, decided to experience for themselves what it was like being admitted into a psychiatric hospital. For the purposes of the experiment they chose several hospitals in five different American states and tried to get admitted by simply appearing on their doorsteps.

In all of the cases the experiment started in a similar way. The pseudo-patients would first call the hospital and set up an appointment to be examined. When they came to the reception they would complain of sometimes hearing voices. When asked what kind of voices these were, they would answer that they were mostly indistinct, but that some words were recognizable, among which "empty" and "hollow" would often appear. These very words were chosen deliberately because there was no report in the medical literature of the time on hallucinations dwelling on emptiness and the meaningless of life. The psychiatric profession had not yet written about the supposed existential psychosis which these "patients" had enacted. The men in the group heard male voices, and the women heard female voices, and in all cases none could link these inner phantoms to any familiar person.

Except for their fake symptoms and the details of their profession, occupation and name they provided on arrival, everything else the group told the doctors was true. All questions concerning their personal history and other medical difficulties were answered truthfully. Once admitted into the hospital, moreover, the pseudo-patients behaved in a completely normal fashion. They no longer heard voices in their heads. They had sanguine conversations with both the patients and the staff and followed all instructions they were given except for swallowing the prescribed medication.

Once you get a diagnosis, it's hard to get rid of it

After being admitted, seven were diagnosed with schizophrenia and, when they were discharged, received the diagnosis of schizophrenia in remission. Of course this did not mean that they were healthy, but simply that they were no longer very ill. One of them was diagnosed with manic depression. Even though the doctors treated them as if they were ill, other patients were quick to discover that there was nothing really wrong with them. The actual psychiatric patients would tell them: "You're not crazy. You're a journalist (or a professor, because they often took notes). You are only here to check the hospital's work."

In several cases the pseudo-patients who were hospitalized experienced some anxiety, since none of them had expected that the hospital would be so eager to keep them. Because most of them had never visited a psychiatrist before, they also worried about being exposed as imposters and scorned. Members of the experiment had also been instructed to try and get themselves discharged. For this they would have to convince their doctors that there was nothing wrong with them. Indeed, except for the initial lie about hearing voices each of them had told in the hospital reception office, they had given staff no grounds for believing they were psychologically disturbed. However, each group member soon found out that the only way they would be released would be if they agreed with the doctors that they also believed they were mentally ill and then declare that they were feeling much better. The length of hospitalization for the fake patients ranged from 7 to 52 days, with an average period of 19 days spent in hospital.

For their research they tried to write down as much of what they saw and heard as possible and, out of fear that the notes would be destroyed by the hospital staff, hid them in safe places each day. At first they made these notes in secret, expecting that they would not be allowed to continue; but when staff showed no particular interest in this activity, they began writing in their journals quite openly in the common social rooms. It turned out that taking notes was not deemed problematic, as hospital staff usually interpreted it as a compulsive disorder causing the patient to write everything down. Once they were labeled as schizophrenic, even the fakes' most habitual and inconsequential activities were viewed as potentially pathological.

During the entire period of the experiment the fake patients were prescribed a total of 2100 pills. Of these, only two were swallowed and all others ended up in pockets or being flushed down the toilet. Another interesting thing is that the fraudulent patients often discovered that the toilet bowl already contained medication left there by actual patients.

Expecting pseudo-patients

During their research the team observed two major common symptoms in people admitted to psychiatric hospitals, notwithstanding their individual illnesses: all suffered from a feeling of helplessness and a feeling of depersonalization. The staff talked about the patients as if they were not present in the same room or even invisible. On one occasion, for example, a nurse unbuttoned her uniform in front of the entire ward and readjusted her bra although many male patients were present. But this was not at all meant as an act of exhibitionism: the nurse simply did not regard the patients as physically and socially real people, in front of whom she would feel normally uncomfortable shifting her underwear.

When doctors from other hospitals across the United States heard of this singular experiment they were naturally certain that nothing of the kind could happen in their wards. Rosenhan therefore arranged with one of the most prestigious hospitals to send them a couple more false patients over the following three months. Senior doctors in the hospital were convinced that they would have no problems in identifying the impostors. For the next few months, each new patient at the hospital was intentionally evaluated as a possible fraud. Out of 193 patients 41 were identified as false and another 42 were placed under suspicion. Of course the doctors were in for another shock, since Rosenhan had not in fact sent them a single false patient.

The BBC repeats the experiment

Recently, the popular BBC science program Horizon repeated the experiment. However, this time the volunteers did not try to get themselves admitted into hospitals; the experiment was carried out as a sort of reality show. They chose five typical psychiatric patients with the most common diagnoses: bipolar disorder, depression, eating disorder, obsessive-compulsive disorder and social anxiety disorder. All these people had been in treatment for quite a while, so they could not be instantly recognized as different to the five volunteers who had never experienced any difficulties with their mental health.

Three experienced psychiatrists, experts on mental disorders, followed the group of ten for one week. Their task was to separate the ill from the healthy by observing them and performing tests. For an experienced diagnostician this should not be too difficult, but it soon turned out that this process of identification was a very demanding challenge. The specialists only succeeded in correctly identifying the two disorders that were most characteristic. The obsessive-compulsive disorder was exposed when they observed the patients cleaning a stable since one of the ten was much more meticulous in washing up thoroughly after the dirty work was done.

The experts were also successful in diagnosing the eating disorder because one of the participants showed a 30% deviation in assessing her actual physical appearance. The patients were told to correct a deformed image of themselves using a computer program so that it would match their real image. Most of the participants in the experiment were able to correct their image more or less successfully: only a formerly anorexic patient, who had already been rehabilitated and now leads a completely normal family life as a mother of three, failed to make a realistic image of her body; she thought she was one third fatter than in reality.

When it came to other diagnoses, the experts unfortunately failed. A once severely depressive patient was actually declared to be the healthiest of the group. It should probably be mentioned that all they had at their disposal while making the diagnoses was footage of the candidates performing the task they were given, and the chance to pose a couple of short questions. The diagnoses therefore had to be made on the basis of very limited information. Nevertheless, there does seem some truth to what philosophers, poets and ordinary people alike have been saying for time out of mind: it is often difficult to tell the sane from the mad, or to say who in the asylum should be patients and who should be doctors.

Most submitted to authority

In 1961 Stanley Milgram, a professor of psychology at Yale, placed a newspaper advertisement calling for volunteers, at a substantial fee, to take part in a study of memory. He chose forty from scores of applicants and invited them to his laboratory. They would only find out at the end of the experiment that the professor was not studying their powers of recollection at all, but was set on observing something very different.

Electroshocks for the good of science

Upon arriving, the volunteers were received by a young scientist in a white gown who explained to them how the experiment would be carried out. The participants were put into pairs and allotted the role of either "teacher" or "learner" at the toss of a coin. Each pair was then taken into a special room where the learner had electrodes attached to his arms. At the touch of a button these delivered an electric shock, initially on a low setting. Both volunteers were given a test shock to give them an idea of the very mild pain the apparatus could inflict. . In the meantime the scientist setting up and observing the experiment reassured them that, although painful, the shocks posed no danger to their health and would cause no long-term damage.

Next the teacher was taken to another room: this contained the device for administering the electrical shocks and an intercom connected to the room in which the learner still sat attached to the electrodes. The teacher was now instructed to ask the learner questions testing his memory. For each wrong answer he was told to punish the learner with an electric shock; for the purpose of the study, he was told, was to investigate the influence of punishment on the subject's powers of recall. However, not all shocks were of the same intensity. Each subsequent shock was 15 volts stronger than the last. From the hardly perceptible 15 volts delivered at the beginning of the experiment the teacher had 30 switches before him, the last of which could send a massive 450 volts through the learner's body. Each switch also had a label describing the danger involved, for example, "moderate shock", or "danger: very intense shock". The last two switches, administering the highest voltage, were simply labeled "XXX".

At first, the experiment proceeded smoothly as the learner answered each question posed by the teacher correctly. Yet as the tasks became more and more difficult, the learner would gradually start to make mistakes. The teacher began administering increasingly powerful shocks for each mistake committed. These were initially still mild, but when the shocks reached 150 volts the learner suddenly did not want to participate anymore. He would start to moan through the intercom, saying that it hurt too much and that he wanted to leave. At 180 volts he would start screaming that he could no longer take the pain. At higher voltages he would whimper and groan, and once the shock level passed 330 volts he would no longer respond. At this point the teacher was instructed that if the learner did not respond in a few seconds this silence should be treated as a wrong answer, and a stronger shock delivered accordingly.

What was this study really about?

The actual subject of Milgram's study was the very question of how the teacher would respond to the learner's suffering. The idea that the project was to research the effect punishment had on a person's memory was only a pretext to get unsuspecting volunteers into the laboratory. Milgrim wanted to see how they would obey the authority represented by the young white-coated doctor conducting the experiment.

This scientist remained in the same room as the teacher throughout the session. He would sit at his desk arranging paperwork while keeping an eye on the subject. When the learner began suffering and protesting, the teacher usually turned to the scientist and asked if it would not be better to call things to a halt. To this the authority figure would at first reply: "Please continue." When the teacher would repeat his request to stop the experiment he would answer: "The experiment requires that you continue." At the third appeal he would say: "It is absolutely essential that you continue." To the final plea he would respond: "You have no other choice, you must go on." It was only when he had given these four, incrementally stronger commands that the doctor would end the experiment – if the teacher still asked to stop.

At this point it was explained to the teacher that the learner had not received real shocks and was not even a real volunteer, but was actually an actor. The increasingly agonized reactions coming through the speaker of the intercom were all pre-recorded. The coin that was flipped to assign the partners' roles had identical sides, so the actor would always become the learner, and the authentic volunteer would always be made teacher. As a finishing touch during the initial electric shock test, the actors would always mention that they suffered from a heart condition, thus planting another moral factor to weigh upon the teachers when they began raising the voltage.

The shocking results of the experiment

Before the experiment, Milgram asked his colleagues how they thought the volunteers would react. For the most part his fellow specialists predicted that only a handful of people with sadistic inclinations would go on administering electrical shocks beyond the point at which the learner starts screaming that he can no longer take the pain. But the actual results of the experiment shocked all those Milgram consulted. 26 % of the forty volunteers displayed complete obedience to the doctor in charge, inflicting the most powerful shocks on their partner in the experiment. Only a third of them showed the will to contradict the doctor's commands and stop the procedure somewhere in the middle.

The experiment was later repeated more than a hundred times in different environments in different parts of the world, but the results were always very similar. About 60% of people were willing to follow the clinical supervisor's commands absolutely, and administer the most intense and deadly shocks. Researchers recently repeated the Milgram experiment for a number of TV documentaries on torture and discovered that the percentage of people prepared to obey an authority's commands to the last has not changed at all in the past decades.

This is not to say that those commands were obeyed blithely or malignly. Milgram reported how participants invariably displayed great distress as they carried out the order to continue, often soaked with sweat, their hands shaking as they operated the switchboard. Many could hardly speak by the time they asked their last few questions. A few began laughing hysterically as the tension proved unbearable. But still they continued to obey. As Milgrim observed of one volunteer:

> A mature and by all appearances cool-blooded businessman stepped into the laboratory confident and smiling. In 20 minutes he transformed into a shivering, stumbling wreck of a man, rapidly approaching the point of a nervous breakdown ... but he continued to respond to every word of the experiment conductor and obeyed to the very end.

Milgram was subsequently accused of having caused permanent damage to the unsuspecting volunteers by subjecting them to such an intense emotional ordeal. Yet when he consulted his volunteers a couple of months after the experiment, to find out whether it had caused them any lasting trauma, Milgram claimed in his report that none complained of any permanent negative effect. Most even expressed no regret at having taken part.

What makes good people do evil things?

The Milgram Experiment, as it became definitively known, was first conducted and discussed in the early 1960s, around the same time as the trial of Adolf Eichmann in Israel. Eichmann was responsible for the deaths of many Jews during the Second World War, but in a Jerusalem court pleaded that he was only obeying orders. Milgram wondered if he could make an average American inflict severe pain on a completely innocent person by creating a situation in which a seemingly legitimate authority demanded atrocious actions in the name of a higher goal.

Milgram later carried out variations of the original experiment to find out how altering its basic elements effected levels of obedience. In one variation, the experiment supervisor would receive a telephone call shortly after the experiment began, and so ask another volunteer (in fact another actor) to take his place while he left the room. Even though his substitute insisted that the teacher continue regardless of the cries from the intercom, only 20 % of the teachers would go through with the most intense electroshocks. Similarly, their willingness to continue decreased in other circumstances where the authority of the supervisor was diminished.

On the other hand, the teachers' compliance with authority increased when they did not operate the switches themselves, but were only one in a team assigned to question the learner. In these circumstances only one tenth of the volunteers decided to quit before the end. However, the personality and the appearance of the supervisor did not affect the results. Whether the experiment was conducted by a young or old man, unkempt or well-groomed, did not make any difference. The teachers' obedience to authority was unchanged by this factor.

Evil, as Hannah Arendt famously said of Eichmann, can take the most banal of forms. The Milgrim Experiment further suggested that acts we call evil in retrospect can arise from our efforts to "be good" and follow the imperative our elders gave us as children – "do as you are told".

Too much safety can be dangerous

The Sunday of September 3, 1967 was a special day for Swedes. From one o'clock at night until six in the morning all road traffic was suspended, except for emergency vehicles, and even these had to follow special regulations. Between 4:30 and 5:00 am, all the roads in Sweden received a drastic makeover. That Sunday, Swedes switched from driving on the left-hand side of the road to the right.

Planning this important change in Sweden had taken several years. In 1955 there was even a referendum in which more than 80% of voters opted against switching sides. However, despite the will of the people, the expert and practical arguments were too convincing to allow Swedes to continue driving "the English way". What contributed most to the decision was the fact that, in all the neighboring countries, people drove on the right lane. In addition, most of the cars in Sweden already had a left-sided steering wheel, so it was just a matter of time before the decision to switch the direction of driving was taken.

Contrary to expectations, the great change of traffic regulations that took place in Sweden on "H-Day" (H for the Swedish *höger*, meaning "right"), as that first Sunday of September 1967 was named, did not result in a higher number of traffic accidents. Quite the opposite, in fact: on the Monday after the new regulations were implemented, the number of accidents recorded was actually lower than the daily average up to that time. It took a whole year until the average number of car crashes rose back to that seen before "H-Day".

The paradox of perceiving safety

In his book *Traffic: Why We Drive the Way We Do and What it Says About Us* (Allen Lane, 2008), one of the questions Tom Vanderbilt asks himself is how an actual increase of danger on the road, like that which took place in Sweden in 1967, can decrease the number of accidents. The answer is, of course, quite simple: upon changing the side of the roads on which they drove, people felt less safe and therefore drove more carefully. After a year's time, when they became used to the new regulations, they became more confident in their driving again and the number of accidents rose back to the previous level.

On the basis of additional examples, Vanderbilt concluded that in perceiving road danger we face a paradox: "The system that the majority perceives to be more dangerous is in fact safer, while the system that seems safer is actually more dangerous."

The explanation of this paradox is that people adjust their road behavior according to the feeling of risk or danger involved. If they perceive something as dangerous they will naturally be more careful than when they feel that nothing bad can happen to them. When, for example, a traffic light turns green, most drivers feel that, at least for a few moments, they own the road ahead and do not need to think about anything but stepping on the gas pedal. If a road is constructed like a racetrack the people using it will accelerate and drive faster even if there are speed limit signs on the side, because such a road makes them feel safe. Similarly, people who drive to work along the same road each day drive much faster than tourists taking the same route for the first time, which is completely understandable, as locals know every detail of the way and therefore feel much safer.

There are several approaches to planning and building roads. According to a theory of "passive security", roads should be built in such a way that potential mistakes made by drivers would not cause severe consequences. This way of building could be called "the engineering approach" as it tries taking into account in advance the potential errors made by drivers, much as builders and architects must consider the possibility of an earthquake or violent winds. The problem with this approach is that it does not take into consideration one major difference between an earthquake and a dangerous driver. The earthquake does not "know" that people have begun building more solid structures, so it will not become more violent than it might have been in the days before "earthquake-proof" buildings. Drivers, however, are less reliable than earthquakes in this respect. They tend to feel much less threatened on "safely" built roads, and so drive less cautiously. This in the end, can result in even more accidents than on roads which seem much more dangerous.

One of the pioneers of a completely new approach to road planning and traffic signalization was the recently deceased Dutchman Hans Monderman, who is also introduced by Vanderbilt in his book. The method Mondermann developed for regulating traffic is the complete opposite of the aforementioned "engineering approach", which designs a road plan around a set of predictable driving mistakes. Monderman, on the other hand, thought of himself as an architect rather than an engineer. For him the problems of traffic regulation involved human factors more than purely physical forces.

Mondermann's theory is best summed up by considering two types of spaces in street and road construction. The first is the "traffic space": this is best represented by the freeway, built exclusively for vehicles and enabling an efficient traffic flow. According to Monderman, the best example of this type of space is the German "autobahn". Typical environments representing the other basic type of space are city and village centers, which he calls "shared spaces". Here, the vehicle is only a "guest" or, in the best case, an equal to other participants in traffic, such as pedestrians, cyclists, children, domestic animals, horses and other non-motorized road-users.

According to Monderman, the crucial mistake committed by classic traffic engineers is to try transferring the rules and standards of the "traffic" space onto the "shared" space. The abundance of traffic signs, road markings and other types of signalization in cities and villages transposes the rules from the world of cars into the world of pedestrians. If the road outside a village is designed in the same way as the road inside the village, only with a speed-limit sign, people will only drive slower if they are afraid of police control. That is why Monderman suggested a radical solution: the shared space in city and village centers should not have any traffic signalization at all.

The road as a space of participation

Monderman first tested his ideas in several Dutch villages and towns. In the village of Oudehaske he tried making the main road as "rural" as possible, and after it was rearranged, tested it by measuring speed. He wanted to find out whether the perception of the road really exerts such a strong influence on drivers. When measuring the speed of cars he was at first convinced that his laser speed gun did not work, but he quickly realized what the problem was. Everyone was driving under 30 km/h and the gun could not detect such low speeds.

In his test village Monderman deliberately did not try to organize the traffic, but intentionally made the conditions it faced more complex by putting pedestrians, cars and cyclists together. Drivers no longer felt comfortable on this road, and as a result had to communicate much more with other people using it.

Reassured by his initial successes, he decided to take on bigger challenges in cities. In the city of Drachten he rearranged an intersection which made him famous around the world. Although it was an important city intersection through which up to twenty thousand vehicles drove each day, along with a large number of pedestrians and cyclists, he removed all traffic lights and signs, and reduced the road markings to a minimum.

However, he did not leave the space entirely empty. He added a series of fountains fitted with sensors which could detect traffic density: the water these spray into the air rises higher when the number of vehicles increases at the intersection. He put a round, green plot in the center of the intersection to create a roundabout of sorts, but this was a roundabout which motor vehicles were not allowed to dominate. All the various kinds of traffic meeting at this island must communicate with one another and negotiate their way through co-operatively. Despite the fact that the intersection also includes a driveway to a shopping mall, the traffic is completely fluent, and even more so than it was before, when traffic lights ruled the intersection.

Videos of traffic at intersections and on roads rearranged by Monderman may be found on YouTube. The thoroughfares reflecting this "psychological" approach to traffic immediately received the name "shared space". Monderman put so much trust in theories to the extent that he once answered a journalist's questions right in the middle of a busy road he had planned, and had no problem with walking through his intersection backwards or with his eyes shut. He knew that, because of the way the road and the intersection were organized, drivers would pay equal attention to all kinds of traffic and would not by annoyed by someone "getting in their way"; in Mondermann's view of traffic flow, we all get in one another's way, and the only way to proceed is for all concerned to be heedful of one another. Drivers on a Mondermann road would thus even adjust to a weirdo walking backwards across it.

What does the peacock's tail say?

A gazelle grazes in the savannah. Suddenly, it notices a dangerous predator approaching through the high grass. At first it stands motionless. Then, instead of bolting, it leaps high in the air and continues jumping about two meters high right in front of the beast hunting it. But why? Is this not a complete waste of energy, and would it not be better to run away as quickly as possible? Does it use jumping to warn other gazelles in the vicinity?

At the beginning of the seventies a biologist called Amotz Zahavi, a professor of zoology at Tel Aviv University, started examining a question that has puzzled scientists from Darwin on: why do animals, in terms of energy consumption, often behave in wasteful ways? Why do peacocks have bushy tails which only hinder their movement through their natural environment? Why do gazelles, at the sight of a lion, not run away, but start to jump high into the air?

Zahavi developed a theory according to which the seemingly irrational and wasteful behavior has no meaning in itself, but functions as a message. With its high leaps the gazelle is telling the lion that it is healthy and strong, so that any attempt to catch it would be useless. The lion would need more energy to catch it than if it decided to look for another animal less agile and fast. Similarly, a peacock with its bushy tail informs potential mates that its genes and physical condition are excellent; otherwise, it would not have been able to grow such a magnificent tail.

Which signals can be trusted?

But why would animals "tell the truth" with these signals? Would it not be most convenient for every living being to boast and show itself in the best light possible in order to survive? Could an animal gain anything at all by "telling the truth"? How can sincere communication be established among living beings in nature when there is a conflict of interests between the sender and the receiver of information?

According to the theory developed by Zahavi, and independently by several economic theorists, the solution to the problem of trustworthy communication lies in the high costs of creating such signals. Zahavi claims that the very fact that animals have to invest such amounts of energy into creating these signals is of key importance. It is exactly because producing these signals in the natural environment seems to be an irrational waste of energy that they can be acknowledged as genuine and not deceptive in the eyes of the receiver. If every signal has a certain price and if the creation of false signals is on average and in the long term more costly than the creation of true signals, it simply does not pay to lie.

Drawing from sports, Zahavi named his hypothesis the handicap principle. The energy-wasting nature of signals functions as a sort of additional difficulty which guarantees that messages are truly genuine and not false. The peacock's tail is the perfect example of such a "handicap" or energy-wasting signal. The purpose of the bushy tail is to communicate to potential female mates what they cannot perceive directly. Maintaining a lustrous, bushy tail requires a great amount of energy, which only strong and healthy males can afford to spare. Sickly and weak peacocks would be so exhausted by such an investment that they would not be able to reap its benefits. They have to direct all of their energy into survival alone and have no extra energy left to invest into luxuries like an extravagant tail.

Another important fact is that peacocks cannot benefit from lying in the long term. Investing in a tail requires such an expenditure of energy that it is better for them to hope to succeed with one of the females without growing a bushy tail. It might be that they are only weak one year and will be able to grow a magnificent year by the next, but for the time being it is of foremost importance to survive until that time. It is because of its high energy cost that a bushy and colorful tail makes a good indicator of a peacock's condition, which would have otherwise remained hidden from females.

Biologists discovered similar communicative signs in birds. Chicks in the nest can be very loud when they are hungry, because their feeding depends on what their parents bring them. The dilemma a bird parent faces is how to know which of the little beaks is really in the greatest need of the worm it had just brought to the nest. Is it the chick that chirps the loudest? If the chicks had no natural enemies in their surroundings, the bird parent could not be sure whether the loudest chick is also the hungriest. As the loud calling can also attract enemies, the chick, once fed, is better off not making any more noise which might cost it dearly in a wilderness full of threats from rivals and predators.

Why it makes sense to tax luxury goods

Similar conclusions to those made by zoologists studying animal behavior in the wilderness were also reached by economists. People also like to invest in goods which are not really necessary for their survival. It was already in the eighteenth century that Adam Smith realized that people are very fond of displaying their wealth, while they tend to conceal their poverty. However, he did not (yet) see displays of wealth as status symbols, but formulated the theory that people prefer to identify with happiness rather than with misery, making it sensible to show off wealth and hide poverty.

A more modern point of view was adopted in the middle of the nineteenth century by John Stuart Mill, who perceived luxury goods as status symbols and suggested that they be taxed. As people do not buy expensive things because they need them, but to show how well they are doing, their price naturally does not correlate with their usefulness. In Mill's opinion that is why it is only proper that the government makes these items even more expensive by imposing taxes.

> When a thing is bought not for its use but for its costliness, cheapness is no recommendation ... a tax on the article is really paid by nobody: it is a creation of public revenue by which nobody loses.

If wealth was once no more than a sign of an individual's success, it gradually became a value in itself. At the end of the nineteenth century, the economist and sociologist Thorstein Veblen analyzed the ways in which people exhibit their wealth in his book *The Theory of the Leisure Class*. At first, the mark of a well-off individual was that he did not have to work in order to survive. Having plenty of spare time to dedicate to leisure activities was a good indicator that a person was wealthy.

As an example of displaying an abundance of spare time he also gave a long "liberal" education, which he contrasted with training aimed at developing the skills required to perform practical tasks. Like proficiency in sports, an extensive education aimed at learning unimportant skills was not much more than an exhibition of a person's abundance of spare time.

On the subject of spare time it should also be mentioned that modern science first developed in these very wealthy circles, because someone trying to earn a living through research could not be thought entirely credible. A trustworthy scientist could only be someone who researched for pleasure in his own spare time, without having to rely on his work to support his family. Scholars who were employed at universities received money for teaching and not for research. This belief remained unchanged until the nineteenth century, when being a scientist became a profession like any other.

What is an education good for?

The theory of education as a signal was perfected by the economist Michael Spence and earned him a Nobel Prize in Economics in 2001. While Veblen defended the claim that higher education was a signal of belonging to the work-free social class, Spense developed a theory according to which the level of education is primarily a signal to potential employers, conveying information about the characteristics of an individual which could not otherwise be shown. In 1973 he published a mathematical model of the employment market based on the level of education as a signal which, much like a peacock's tail, expresses the otherwise unperceivable characteristics of an individual.

In a simplified version of this model there are two sorts of people on the employment market: in group A there are highly productive workers, and in group B there are less productive workers. However, an employer cannot evaluate in advance whether the potential employee is a group A or a group B worker. In his original model Spence does not take into consideration the possibility that additional education would directly increase the worker's productivity. The essential question is why would potential workers in this model educate themselves at all, and why would employers give higher salaries to more educated people?

Spence demonstrated through his analysis that a person's level of education can be a good indicator of the type of worker they are. Much like the case of the peacock's tail, a degree or a doctorate which requires a considerable effort and the investment of a great amount of energy, on average gives a credible signal of an individual's potential productivity. However, it should be pointed out once again that, at least in this simple model, education has no practical significance and in itself does not increase the productivity of an individual. It is only a signal aimed at the employer, which is in most cases proves trustworthy. For the less productive group B, there is no sense investing in additional education in hope of a higher salary, since they would have to invest more in acquiring qualifications than they could gain by having them. That is why education, according to this model of the employment market, is as trustworthy a signal for employers as the high leaping of a gazelle in front of a lion or the peacock's proud display of its magnificent tail.[1]

---

[1] Carl T. Bergstrom, "The Theory of Honest Signaling," http://octavia.zoology.washington.edu/handicap/.

Who really makes the decisions in our minds?

Researchers invite you to take part in a memory experiment. Your task is simple. You have to remember a number and transfer it from one end of the building to the other. You are given your task in the reception room, then you go down the hall towards the other end of the building where you have to communicate the number you had been told. While you are walking down the hall, struggling mentally not to forget the number, one of the experiment organizers suddenly stops you and, for refreshment, offers you a choice between a tempting chocolate cupcake and a low-calorie fruit salad. You choose, and then continue on your way to the designated room.

It is only after you have successfully passed on the message that the researchers tell you they were not even interested in your ability to remember the number, but in how you chose your snack. The main purpose of the experiment was to find out whether you would choose the cake or the fruit. The participants in the experiment were not all given equally difficult tasks. Some had to remember a seven-digit number while others were only given a two-digit number. Strikingly, a larger percentage of those who had been given the more demanding task chose cake instead of fruit salad. 59% of all those carrying a seven-figure number in their minds chose the cupcake, compared to only 37% of those who were asked to pass on a two-figure number.

Do people really make rational decisions?

This and many similar experiments to observe people's decision-making processes in different circumstances indicate that our choices are not nearly as autonomous as they may seem. Since the early Greeks, humans have preferred to believe in the existence of free will. We aspire to an idea of ourselves as are essentially rational beings, making decisions on the basis of careful consideration, supported by logical analysis. Accompanying this ideal, however, is a perennial sense that we are governed by more powerful forces both beyond and within ourselves. Reason is not only prey to circumstance but is also clouded by the passions, and by the frailties and desires of the body. The latest research on how we make our choices casts compelling light on these age-old preoccupations.

Let us look at another experiment conducted recently. In his book, *Predictably Irrational: The Hidden Forces That Shape Our Decisions* (Harper, 2008), the American economist Dan Ariely analyzes an apparently absurd offer for a subscription to *The Economist* magazine. Potential readers were given three ways of subscribing: they could receive the digital edition of the magazine at a discounted rate, or the printed edition at a higher price, or take a third package at the same cost as the printed edition, but giving them both the printed and the digital editions together.

Naturally, nobody in the test group decided to subscribe only to the printed edition, because they could get the digital edition as well at the same price. So why did the advertisement even offer the printed edition only when, for obvious reasons, nobody would choose it? The answer to this question is hidden in the results of the second part of the experiment in which potential subscribers were not offered the second option and could only decide between the digital edition and the combined subscription. Of 100 students who took part in the first part of the experiment 16 chose the digital subscription, 84 chose the double subscription and none took the printed edition only. However, when just two options were available, 68 participants chose the cheaper digital edition while only 32 decided to go for the more expensive double edition.

By floating the offer of the printed edition alone, a manifestly poor purchase, the advertisers enticed a substantially greater number of people decided to make a choice which cost more. . The offer for the printed edition by itself was a ruse, deliberately employed to create confusion in customers' brains. Instead of simply presenting them with a cheaper and a more expensive option, the advert gave readers a decoy which made this pricier deal seem like a real bargain.

The reason for the different result in each of the two experiments is that the brain does not usually assess the absolute values of things being offered, but decides on the basis of their relative values. The brain only wants to know how one offer compares to other available options.

In fact, the brain is very bad at attributing absolute values to things. It is much more successful in comparing the values of individual offers and this is where it can be most easily fooled. That is also the reason why restaurants usually offer a very expensive dish on their menus that few would be likely to order. The purpose of this dish is to make the visitors of the restaurants feel better when they order something cheaper, because they then feel they have got a good deal.

Which part of the brain can be trusted?

Scientists are learning that our brains have two separate systems for making decisions. The first is specifically human and works by analyzing a given situation rationally. With this part of the brain we weigh up all available options until we have found the best. The brain's other decision-making system, meanwhile, has been inherited from our animal predecessors and works on a subconscious level. Even as your brain carefully analyzes the options before it and scans through the smallest of details, you are not really aware of this subconscious process, but simply feel that a certain choice is objectively more attractive than the others.

This second system of decision-making informs us of its preferences through feelings or emotions that affect us when we are considering an individual offer. If an option is extremely unfavorable we can feel it in our bellies that we should decide against it. Nevertheless, both decision-making systems have their good and their bad sides. The secret of deciding correctly is to know when to rely on your mind and when to follow your feelings.

A Dutch scientist called Ap Dijksterhuis observed how people made decisions when buying furniture at an Ikea store. He found out that the more time the customers spent deciding what to buy and analyzing their options, the less happy they were with what they bought later on. In a report on this unusual research published in *Science* magazine a couple of years ago, he wrote that when it came to complex decisions that involved analyzing a great amount of data, people made much better choices when they relied on their instincts. It was only with simpler choices that a logical analysis turned out to be the better strategy.

The latest research suggests that to make good decisions one should do exactly the opposite of what tradition or common sense tell us. Conventional wisdom teaches that when dealing with simple decisions, we should not rack our brains about making the right choice, but trust our instincts, while with more difficult decisions, we should put our minds to it and decide on the basis of a thorough analysis. Yet Dijksterhuis and other scientists came to the conclusion that the best strategy was exactly the opposite.

The problem with the analytical part of our brains is that it can only process a limited amount of information at once. According to various measurements, a human being can rationally process only five to ten independent bits of data at the same time. When the amount of information exceeds this number, the analysis becomes unreliable, as the brain begins to generalize, which can cause an important piece of data to get lost. When there is too much information, a person trying to decide on the grounds of rational analysis often makes a wrong or less than ideal decision.

In the first experiment described earlier, people who had to pass on a seven-digit number were more inclined to choose the cupcake because the rational part of their brains was occupied with remembering the number, so they left the decision to their feelings which made them pick the cake. The participants who only had to carry two numbers in their minds were left with enough space in their "brain processors" to think rationally about the options and chose the healthier fruit salad.

Ap Dijksterhuis believes that the transition from the rational to the instinctive system of decision-making can already occur when we are dealing with a problem that has more than four different variables. The general rule that his research has lead him to is: when making decisions that do not require you to take a large number of variables into consideration, decide rationally; but when it comes to complex decisions, trust your instincts.

With difficult decisions like buying a new car, an apartment or furniture, the best strategy Dijksterhuis suggests is to first collect as much information as possible, then let the brain process it for a while. It is best to leave the decision-making for at least one day and sleep on it. Even if we are relying on our instincts, it is now much more likely that we will choose correctly and be happier with our choice in the long term.

Mirror neurons

In 1996, three Italian scientists at the University of Parma studied the reactions of monkey brains indifferent activities. They wanted to learn, for example, which parts of the brain were active when a monkey reached for food. They discovered that when an animal reached for raisins set on a table, specific neurons were triggered in its brain. These findings were completely routine and had already been described several times before. In the course of their experiment the Italian scientists did however make a very interesting and important discovery by chance.

Discovering mirror neurons in monkeys

In repeating the monkey experiment a number of times, the scientists noticed something that was, at first sight, most unusual. One of the researchers reached for a raisin and ate it himself, right in front of the monkey. To everybody's surprise the sensors following activity in the monkey's brain detected the same neural response made when the animal itself reached took a raisin. This could only mean that the neurons reacted the same way when the monkey both observed the action and performed it for itself.

This unusual phenomenon immediately caught the researchers' attention. They examined it in detail and published their observations in 1996. Since the brain reacted the same way both in performing and observing an identical act, cells responsible for the phenomenon were named *mirror neurons*.

Mirror neurons in humans

Similar experiments were soon carried out in other laboratories, and researchers quickly discovered that mirror neurons were triggered not only when a monkey observed an action it could then repeat, but also when it heard sounds accompanying a familiar situation.

Scientists then began conducting similar experiments on mirror neurons in people. They quickly found out that in humans the phenomenon was even more evident. In 1998 researchers discovered that a great concentration of mirror neurons in humans was to be found in the brain's language processing area. In a number of recent experiments it was found that our brains react in a similar way whether we listen to an action being verbally described, watch it being performed, or do it ourselves. The mirror reaction in monkeys extends to when they see or even hear an act being performed which they can also accomplish for themselves. Mirror symmetry in humans, however, is more sophisticated and wide-ranging, extending to when an action is verbally described to us. When someone listens to or reads a narrative their mirror neurons respond as they would if that person were actually doing the act or experiencing the event being related.

Mirror neurons, from nature to culture

On the basis of these surprising findings scientists hypothesized that the significant increase in the number and function of these mirror neurons in the brain is one of the crucial evolutionary developments that enabled humans to make their incredibly swift transition into culture some tens of thousands of years ago.

Mirror neurons enable a highly efficient transfer of knowledge. The ancestors of the human species were also quite resourceful, but their mirror neuron mechanism was probably not sufficiently advanced for the quick and efficient transfer and distribution of ideas. Among such forebears new ideas were destined to disappear quickly because they could not store or pass on such information very efficiently.

One of the hypotheses circulating among researchers of brain activity is that autism might be one consequence of irregularly functioning or even inactive mirror neurons. One leading expert in this area, Vilayanur S. Ramachandran, professor at University of California, San Diego, looked for empirical evidence to support this theory. Examining specific parts of the brain, he measured the level of mirror neuron activity in healthy individuals and patients suffering from autism. He set the simplest kind of mirror neuron test: subjects were asked to perform a task and also watch someone else doing it.

With healthy people he found the brain activity characteristic of normal mirror neuron function: the neural response, that is, was similar in performing as well as in watching the set task. I In autistic people the response was much weaker. Since mirror neurons apparently govern our ability to imitate and empathize, scientists now believe that this deficiency in the brains of those suffering from autism is responsible for the difficulty they often have interacting with other people and to emerge from an inner mental universe.

During recent years, research on mirror neuron function has become much more far-reaching. That is why Ramachandran's claim from several years before is no longer as surprising as it might have been at first. In his opinion, mirror neurons will do for psychology what DNA did for biology: with their help we will be able to develop a unified model to explain many mental capacities that would otherwise have remained beyond the reach of experimental research.

Brain plasticity

Researchers of the brain have long believed that once fully grown, it no longer changes. Only the young brain was thought to be "plastic", as the ability to adapt is called in technical terms, and it was believed to lose this plasticity with time. However, research has shown that this theory is not completely correct.

Mapping monkey brains

When scientists first began to investigate in detail which areas of the brain control movement in individual parts of the body, at the beginning of the 20th century, they found that their results were not alike in all the monkeys they tested. The researchers thought at first that the discrepancies might be due to a lack of precision in their experiments, but it soon turned out that, although mapping brain functions was by no means a simple task, their puzzling findings were not caused by methodological errors.

Researchers wanted to find which area of the brain controls movement in various parts a monkey's body. To do this they stimulated one region of the brain after another, moving methodically through specific points within each region and recording physical movements as they occurred. . By stimulating one point, for example, they triggered the movement of a finger; another point caused the entire hand to move. Looking back, it is hard not to reflect upon the monkey's side of this purely mechanical investigation. As brain tissue is not susceptible to pain, the procedure was not painful for the animals, but it cannot have been very comfortable either.

More accurate measurements have since confirmed that these early brain-maps were correct in identifying discrepancies between individual brains. The point that triggered, for example, movement of the hand was not located in exactly the same place in all test subjects. It turned out that the maps were to some extent specific to every individual monkey, much as fingerprints are unique to every individual person.

The maps also revealed that the movements these animals performed together, in sequence, were controlled by neurons (brain cells) located close together. Scientists also discovered that the tissue directing the movements most characteristic of monkeys in general took up a larger proportion of the brain than tissue coordinating less typically "monkey-like" actions. If we were to illustrate this with an example from the human world, this would mean that the brain of a violinist has a substantially larger and more developed area controlling the fingers of the left hand than someone who has never played a musical instrument, an activity requiring great dexterity. Similarly, professional dancers possess a much larger area responsible for foot movement than people who only use their feet for walking.

Do brain maps change?

A question that soon arose from this research was how these areas in the brain were formed and parceled out; and consequently whether they could change during an individual's lifetime. The more or less generally accepted belief of scientists for the greater part of the 20th century was that these areas took shape in early youth and could not undergo significant changes later. However, some skeptics decided to try and find out for themselves whether it was really true that the map of a monkey's brain would not change with time.

It was during the period between the two world wars that scientists discovered that brain maps do indeed change, most likely in proportion to the use of individual muscles. The movements executed more frequently were represented to a greater extent within the brain than less frequent movements. However, these experiments went more or less unnoticed because of the already established view that a fully grown brain could not undergo any further structural changes.

Despite this consensus, in the seventies American neurologist Michael Merzenich and his colleagues decided to investigate more thoroughly how brain structure might change under different external influences. At first, Merzenich wanted to find out what effect there would be on a monkey's brain if sensory information was cut from a specific part of its arm. He first accurately determined which areas of the monkey's brain were responsible for processing the sensory stimuli from specific parts of the body. He then performed an operation that caused the monkey to lose feeling in one of its palms. After some time he mapped the monkey's brain again and discovered that the area which had previously been responsible for the sensations in the thumb and its surroundings was not blank, but was now processing information from another part of the arm which still sent signals to the brain, because the nerves from that part of the arm had not been disconnected.

After this discovery he directed his research into less invasive experiments. He was curious whether the brain maps would also change if he taught the monkeys a new skill. He and his colleagues conducted a rather elaborate experiment in which they taught adult monkeys to use their fingers with great precision. The monkeys would get a reward if they used the correct amount of pressure while handling a device. For them this was not at all a simple task, because they had to invest a lot of effort into training their fingers in order to win their prize. It was no surprise that at the same time the researchers also discovered that the area in the monkey's brain which controlled the movement of the fingers grew substantially with use.

How can the blind read?

The same characteristics that had been discovered in monkeys were later confirmed in humans. One interesting piece of research focused on how the brains of blind people could process Braille. Scientists first discovered that the area of the brain which processes the sensations in the reading finger of a blind person grows with practice. This was hardly surprising; what was more fascinating, though, was that the area responsible grows at the expense of the surrounding region, which interprets sensation in the Braille reader's other fingers. This adjacent tissue diminishes as the "reading" cells become more developed.

Researchers stumbled on an even greater surprise when, using more modern methods, they discovered that during sleep, even in blind people, the area which is otherwise responsible for processing visual signals becomes active. They expected this part of the brain to be dormant in blind people, as they do not receive any signals from their eyes. In their dreams, however, it would seem that the blind still see; or still at least use the visual dimension of their minds.

Moreover, it also seems the brain is so adaptable or, technically speaking, plastic, that an area not being used adjusts and becomes specialized in other tasks. As recently as the mid-nineties, however, this principle would not have been widely accepted. The scientists who observed the brain activity of blind people reading Braille encountered great difficulties publishing their findings. *Science* magazine refused to print their work, because the editorial staff could not come to terms with how completely different parts of the brain could interconnect and work together, performing new tasks. In time, the article was published in the rival journal, *Nature*.

Later, new research revealed that the visual part of the brain in blind people is in fact maintained by the essential role it assumes in the fluent reading of Braille. Someone who can read Braille really well does not feel the dots on the paper at all, but directly perceives entire words, just as someone reading ordinary writing is not conscious of individual letters, but takes in words and sentences as visual units.

In the year 2000 there was even a case described in scientific literature of a woman who had been blind since early childhood and had become an extremely accomplished Braille reader at school. When she was 26, though, she suffered a stroke in the area otherwise responsible for processing visual information. It would seem reasonable that such a stroke should not have significant consequences for a blind person, but this case strongly suggested the opposite. For although the woman could still feel the dots of Braille under her fingers, she suddenly no longer understood their meaning. Her ability to read Braille was stored in the area that normally processes visual information and which had had been damaged by her stroke.

Mental fitness

It has only recently become clear that even the older brain is much more plastic or adaptable than it was first presumed to be. Nevertheless, the fact does remain that the ability to adapt diminishes with time. Younger brains are simply much more plastic than older brains.

Michael Merzenich and his team, already mentioned pioneers in researching brain plasticity, set up two companies offering professional help in overcoming difficulties in brain function. *Scientific Learning* helps children with learning disabilities. Using special computer games which are designed to develop specific parts of the brain that cause these children to have problems, the company has achieved impressive results, success which has more recently led the group to create their *Posit Science* company, an organization focused on helping the older population. This firm has developed various "mental fitness" techniques to aid older people keep their brains in trim, and go some way to recovering their mental prime.

The invention of permanent innovation

Many scholars have tried to determine the causes of the rapid development we have been able to witness in society during the past two centuries. Even though history has seen several earlier periods during which humanity progressed relatively quickly, these intervals were mostly short and restricted to certain areas of activity. Since 1800 however, our civilization has been continually developing at a pace much faster than ever before. A number of hypotheses have been made to explain this rapid development. Today, the most convincing theory seems to be that the Age of Enlightenment brought about an important shift in society's attitude towards new ideas.

What caused permanent economic growth?

In the fifties, the American economist Robert Solow researched the causes of rapid economic growth. Before, economists were convinced that you could produce more if you had more capital and a larger workforce. In simpler terms: if you want to manufacture more cars, you need a bigger factory and a greater number of workers. However, using mathematical models, Solow was able to calculate that increasing the quantity of capital and workforce could only account for about half of the long-term growth of the standard of living. The key question was what caused the other half of growth.

It turned out that the missing half could be attributed to technological innovation. When artificial fertilizers were introduced in agriculture, for example, the same area of land could yield much more wheat with the same amount of work. A similar trend emerged in industry where constant technological improvements enabled the same number of workers to manufacture a greater quantity of products to a higher standard. Solow's finding caused a paradigmatic shift in our understanding of economic growth and encouraged further research into the development of technology and the mechanisms inspiring innovation within society. Solow's work on determining the causes of economic growth was rewarded with the Nobel Prize in Economic Sciences in 1987.

Economic growth occurs when people transform goods to increase their value. A useful example explaining the phenomenon of economic growth is that of a kitchen in which, using the recipes we know or find in cookbooks, we mix different ingredients to prepare meals that have a higher value than their constituent ingredients. A kitchen can be seen as something combining capital (kitchen applications and unprocessed food), a workforce and the various processes of food preparation described in the recipes being used.

One of the more important conclusions of economic sciences in the 20th century is that economic growth does not only result from a perfected or intense cooking process that we get more and more familiar with, but also from new, better recipes and newly acquired cooking skills.

One of the essential discoveries made in Europe to bring about permanent economic growth and a continuous improvement in the standard of living was the establishment of institutions encouraging and supporting innovation. Craftsmen traditionally concealed the secrets of their trade which made the exchange of ideas extremely difficult. It was not until the rights of the innovator became legally protected that new ideas appeared in public and, in turn, began evolving much more rapidly.

Ideas never run out

That is why, in essence, the industrial revolution was in fact a revolution in the social attitude towards innovation. It was then that a system of improving technology and practices was established, leading to the democratization of the means for developing new ideas. Before, those who wanted to be innovative had to be wealthy enough to pursue their activities for pleasure or were doing so in the name of their benefactors. A few centuries ago people started to make money with their innovations while ideas began circulating which was crucial for the process of innovation.

In the 1990s the American economist Paul Romer reached the conclusion that even Solow's model was inadequate. Solow identified the importance of technological innovation to our understanding of economic growth; but he did not explain what made technological innovation possible, or what wider effect it then had on society. Romer therefore supplemented the model by introducing the idea of the effect innovative companies have on the dissemination of knowledge in their surroundings.

Whenever a company tries keeping its innovations secret, eventually at least a part of the technical information leaks out to the public, if not before, then certainly when the company introduces its product to the market. This benefits other companies in similar fields. A piece of innovation, once people understand it, engenders the knowledge and skills that make further innovation possible. The emergence of a new idea in a given milieu stimulates a wider generation of further fresh ideas.

When employees in different companies socialize and exchange bits of business information in an informal setting, this encourages the circulation of ideas. Economists call this an overflow of ideas from the "inner circle" where they originated to the entire system. That is why establishing environments where people may interact personally is very important, be it in a university campus or a technological park where they might meet to share a meal.

That is also why the concept of Silicon Valley, the home of many of the key companies in the field of information technologies, simply cannot be transferred elsewhere. Creating an appropriate microclimate for innovation involves moving a sufficient number of people and establishing an adequate infrastructure. When a country decides to create new "centers of innovation" these must be located close together so that people can meet, get to know each other and exchange ideas. This is the only way to ensure that such centers have a long-term effect on society and become the real force for economic growth which policy-makers expect them to be.

Meta-ideas for encouraging ideas

According to Romer, there is a substantial difference between capital, work and ideas. The same capital or workforce cannot be used by two individuals at once unless they are divided or shared. This does not apply to ideas. Many people can use the same idea without ever exhausting it. Economists tend to say that innovation knows no rivalry. If two people exchange an object, each of them will come and leave with only one thing. If they exchange ideas, they will both leave with two ideas.

This gives great importance to meta-ideas, or ideas about how to encourage the creation and circulation of new ideas. With this in mind the English invented patents and intellectual property in the 17th century. In the middle of the 20th century, Americans established a system of tenders for financing core scientific research, enabling scientists themselves to decide which research proposals merit the necessary funding and which do not. All of these are institutional innovations which make sure that as many ideas as possible come to life.

Romer is also convinced that it is impossible to say what might be the next great new meta-idea that could further improve the efficiency of idea production. However, he predicts that the leading economy of the 21st century will be the one that succeeds in finding a way to encourage greater innovation within the private sector.

An example of an initiative for boosting innovation is already being tested by successful companies like Google. Their engineers are encouraged to dedicate one fifth of their working hours to resolving an issue which is not part of their regular work, but which they find to be of particular importance. This policy has supposedly led to many groundbreaking inventions that are now an integral part of the company's appeal to their consumers, inventions developed by engineers during these "free office hours".

Similarly, the problem of poor countries might not only be the lack of capital and goods, but most of all the shortage of ideas or the absence of an environment in which new ideas would be appreciated. If a poor country decides to increase systematically the level of its citizens' education and encourage their people to apply ideas that have already been proven to work around the world, such a country will start to develop at a rapid pace. Swift economic growth in countries like China is in great part due to adopting ideas developed and tested elsewhere.

According to Romer, an environment hostile to new ideas is one of the biggest obstacles preventing some parts of the world from wrestling their way out of poverty. A few years ago Romer left his post as a professor of economics at Stanford University and started to lobby for the establishment of new urban centers in the underdeveloped world (chartercities.org). These entirely restructured cities would have all the institutions of the modern developed world, while bypassing the corruption, bureaucracy and poor legal protection which hinder progress in the underdeveloped parts of the world. In some African countries, for example, it takes more than a year to start a company. New urban centers would enable people to realize their business plans quickly and securely, without subjecting them to the whims of corrupted local officials. One of the ideas is to establish a new "Hongkong" in the area of today's Guantanamo after Americans shut down their military base.

Are rewards bad for innovation?

Soon after the Second World War, American researcher Harry Harlow laid a trail which would lead to important discoveries about the nature of creativity, but unfortunately his work did not interest anyone for many years. At the University of Wisconsin, Harlow founded one of the first laboratories for studying the behavior of primates. In 1949 he conducted an experiment in which monkeys had to learn how to solve a simple mechanical problem. He gave the monkeys mechanical toys that required them to pull out a bolt to release a lock on the toy.

Most of all, Harlow was astonished to see how the monkeys learned and improved r at solving the problem even though they were not rewarded for their efforts. What he found to be particularly strange was that the monkeys made more mistakes when they were given candy for their solutions. When solving the puzzle, the monkeys were not motivated by an inner desire for food, drinks or the satisfaction of any other biological need, neither were they stimulated by other incentives such as praise or comforting. It seemed that a completely new kind of internal motivation was at work, one that stemmed neither from basic biological needs nor from those forms of reward and punishment traditionally classified under "the carrot-and-stick approach".

Researching the mechanisms of motivation

Unfortunately, Harlow did not continue researching the origins of this then unknown motivation, but instead began studying the emotional attachment of infant monkeys to their parents – in an experiment which later made him famous. It was only in 1969 that psychologist Edward Deci investigated the mechanisms behind the new form of motivation Harlow had observed years before. However, Deci chose to use human subjects rather than monkeys in his tests. He asked students to solve puzzles, such as one requiring them to assemble a cube from small parts.

The experiment lasted three successive days. One by one, students would enter a room where they were given their task in the presence of the experiment supervisor. Before they had finished the supervisor would apologize, saying that he had to leave the room for a moment. However, the researcher did not actually have other things to do, but simply went next door and, for exactly eight minutes, observed what the students were doing while they were left alone. He found out that they would continue working on the puzzles for an average of another four minutes before they shifted their attention to something else. Among other distractions, magazines such as Time, The New Yorker and Playboy were also put on the table.

When the students returned the following day, half were promised a cash reward for successfully completing their tasks, while the other half or the control group were not told about the reward. As expected, there was a difference between the two groups as the students who were going for the prize turned out to have more success at solving the puzzles at solution. On the third day this group was told that, unfortunately, the researchers had run out of money, so they would not be able to give them a prize that day. Meanwhile, the control group kept working at and solving the puzzles as steadily as before.

It was on the very last day that the greatest difference between the two groups emerged. The group that had been rewarded the previous day showed a significant decline in their motivation for solving the puzzles while the control group, compared to the first day, surprisingly seemed to become even slightly more eager to continue with the experiment. That is how Deci determined that the promise of a reward only caused a short-term increase in motivation, which then dropped rapidly and actually decreased motivation levels for continuing unless another, larger reward was offered.

Scientists travel to India

A group of economists, including one of the leading American-Israeli experimental economists Dan Ariely, recently travelled from America to India to study in greater detail the links between different forms of motivation and productivity. As living expenses in India are significantly lower than in the US, they needed only a relatively small amount of money to set up an experiment in which Indian test subjects would be offered quite substantial rewards.

The group recruited 87 people and asked them to play a number of different logic games involving attention, memory, concentration or creativity. The participants' success was then rewarded with three different prize amounts, corresponding to the average daily, fortnightly and five-month salaries in India. They wanted to find out whether the size of the reward on offer would effect the rates of problem solving.

The results were surprising. The participants who had been promised the medium reward performed no better than those who would only receive the smaller reward. The group that did worst was actually the one offered the biggest prize.

Similar results were later obtained in a study involving students from the Massachusetts Institute of Technology. A first group of students was offered a reward of 600 $ while another group was only promised 60 $. The students then had to compete in two games. The first required mental agility (arithmetic) while the other involved motor skills (pushing a button as fast as possible). It transpired that when it came to strictly physical activities, the higher reward did increase the candidates' success rate. On the other hand, when students had to face intellectual challenges the results were the same as in the Indian study: the higher reward meant less success in completing the tasks.

The evolution of motivation

In his resounding work *Drive: The Surprising Truth About What Motivates Us* (Riverhead, 2009), Daniel H. Pink labeled the instinct to survive or satisfy one's essential needs, such as obtaining food or avoiding physical danger, *motivation 1.0*. This is the basic biological level of motivation common to all living beings. When people started living in larger communities, another form of motivation was developed, based on rewards and punishment. Pink calls motivating by rewarding the good and punishing the bad *motivation 2.0*. This type of motivation gradually became the generally accepted form of increasing productivity in all areas of labor. However, on the basis of scientific research Pink has come to believe that, unfortunately, motivation 2.0 is not the right method to encourage innovation. So what is the problem?

"Algorithmic" forms of work require us to follow an already known and established set of activities which can be described by algorithms. We conform to these algorithms by performing individual chunks of labor and solving more or less predictable problems. In the past most jobs were closely bound to algorithmic labor; today, however, more and more professions require what Pink calls an "heuristic" approach to work. This type of work is not based on pre-established algorithms which dictate how something should be done. Instead innovation is of key importance. When doing heuristic work, we need to come up with new solutions ourselves rather than following more or less fixed procedures to a pre-defined goal, all of which can be neatly summed up by an algorithm. Such work is mostly routine, while the essence of heuristic work is that it involves going beyond a preset pattern. The basic problem of motivation 2.0 is that it does not increase productivity in heuristic types of work, even though it is an effective motivator when it comes to algorithmic labor.

The third form of motivation defined by Pink, the one which is of vital importance in creative activities, is *motivation 3.0*. It is no secret that people are naturally inclined to search for something new, to learn and research. However, this inner impulse is very frail. It is much weaker than the primary motivation to survive and the secondary motivation driven by rewards and punishment. In order for this third form of motivation to prevail, the conditions within the environment have to be sufficiently stable for people not to feel overwhelmed by having to satisfy their more fundamental needs.

The essence of motivation 3.0 is that it stems from an individual's inner desire to create and to be good at something. Unfortunately, motivation 3.0 can be easily suppressed without proper care; however, it can also become extraordinarily productive when nurtured. In his book Daniel Pink highlights three key parts of motivation 3.0: independence, the desire to progress, and doing meaningful work. A creative work environment also demands suitable compensation, so that people do not have think about money and rewards, the essential parts of motivation 2.0.

The fear of punishment and the satisfaction linked to potential rewards, the core of motivation 2.0, influence our brain by limiting its attention when we know precisely what needs to be done. In this case, the carrot-and-stick method increases success rates and productivity at least for a short period. Unfortunately, motivation 2.0 restricts inventiveness as creative work absolutely requires us to be as open as possible to the remotest new ideas and thoughts that might surface in our minds. The pressure of a reward or of punishment can easily extinguish these very subtle flashes that are the essence of innovation.

Einstein and Freud: the meeting of two universes of knowledge

They only met once. It was during the New Year holidays in 1927 when Albert Einstein invited Sigmund Freud, who happened to be in Berlin, for a visit. At the time both were, in their own respects, already living legends. Einstein was approaching his fiftieth birthday and was perceived by the general public as the symbol of the scientific endeavor which at the beginning of the 20th century, had begun discovering that nature was much more mysterious and unusual than anyone could have ever imagined. Similarly Freud, who was more than seventy years old, was regarded as an immense authority in the field of humanities and social sciences. Even those who disagreed with his groundbreaking theories had no difficulties with using his newly established terminology.

When a few weeks after the meeting took place an acquaintance asked Einstein whether he would consider undergoing psychoanalysis, he responded: "I regret that I cannot accede to your request, because I should like very much to remain in the darkness of not having been analyzed." Freud was similarly reserved in his comments on meeting the great physicist. In a letter to a friend, he wrote: "He understands as much about psychology as I do about physics, so we had a very pleasant talk."

Even though at first it might seem that the revolutionary ideas of these pre-eminent scientists are so far-removed that they are alike only in their mutually immense significance to the thought of their time, there are still comparisons to be drawn between them.

Einstein bends time and space

With his theory of relativity Einstein turned upside down centuries of knowledge about nature, space, time, matter and energy.

If, up to this point, science had believed that time and space provided an external framework in which events took place, Einstein's revolutionary theory gave these notions a completely different meaning. Before, it was believed that events that took place in the world had no influence on the course of time or the nature of space. Even if there was something unusual going on somewhere in the universe, time was thought to run unhindered, in a steady flow, without submitting to any kind of disturbance.

After Einstein's revolution, time and space were no longer considered the mere passive background to events that taking place, but became responsive - changing and adapting to everything that happened in the world. In more technical terms, time and space bend according to the distribution of matter and energy. If energy is densely gathered in one place, time and space bend, and can in extreme conditions even "tear". When this happens, a black hole is created.

The scientific revolution brought about by Einstein's theory of relativity is best summarized as a gigantic shift from the assumption that time and space were independent external parameters with which events might be measured and described, to the understanding that time and space are actually influenced by these phenomena. We now no longer see time and space as being impassive to or unaffected by things that happen in the universe, but instead respond to those occurrences by bending or changing form. Einstein's principal achievement lay in describing this bending of time and space with mathematical equations which could predict, for example, exactly how a ray of light from a distant star would change the direction of its movement as it passed the sun. The sun bends the space in its vicinity, which means that light no longer travels in a completely straight line, but only in as straight a line as bent space will allow.

Much as is impossible, despite all efforts, to drive in a straight line on an uneven road, rays of light travel through bent space-time along the straightest line they may. This means that they slightly deviate from their trajectories as they meet dents and bumps in the space they pass through. One of the more prominent curvatures or dents in the space-time continuum that surrounds us is created by the sun. This fact enabled Einstein, applying his equations, to calculate just how much the bent trajectory of light would cause stars to appear to move during a solar eclipse.

According to the theory of general relativity, stars with light rays that passed near the sun would appear to have been slightly shifted because their light had been curved by its gravitational field. This effect is noticeable only during eclipses, since otherwise the sun's brightness obscures the affected stars. The theory was first borne out when Arthur Eddington observed precisely this movement in the stars during a solar eclipse in 1919. The shift, which Eddington measured from a small island off the west coast of Africa, exactly confirmed Einstein's predictions, and subsequently made *him* a star.

Freud and the curvature of mental space

What could be farther away from calculating the bending of time and space in the physical universe than delving into the unconscious structures of the psyche? Yet where Einstein understood events as altering the tissue of space and time, Freud realized how experience changes the fabric of consciousness. The work for which he made his name began with his observing anomalies in human thought and behavior which had hitherto either been dismissed or treated superstitiously. Troubling dreams, strange slips of the tongue, perplexing associations or reactions to seemingly innocuous ideas: these had previously been seen as minor ticks or disorders. To Freud, however, they suggested an entire dimension of the mind and the key to much mental illness. Much as light bends, in Einsteinian theory, around an invisible curvature in space created by the mass of an object; so too Freud observed human thought and expression warping and diverting around hidden anomalies in the psyche. Dreams, bizarre aversions and slips in speech were the expression of things distorted by bends in mental space. To Freud these bends in fact indicated the existence of an invisible region of the mind – the "unconscious" that he is credited with discovering.

Despite this analogy between their endeavors, the problem Freud faced was more or less the inverse of Einstein's. Whereas Einstein had to prove that the curvature of space-time created by mass and energy really was there. Freud, however, had to discover the powerful objects and events in the space-time of the psyche which were creating the anomalies his patients reported. Freud, so to speak, could see the light curve; what he missed was the planet or sun in the unconscious which was causing it. The seminal technique he developed to bring these psychological bulges and rifts to light was psychoanalysis. But psychoanalysis gave the practitioner more than a power of observation alone. In cases where these bends caused suffering, psychoanalysis could be used to modify the layout of a patient's mental space.

Freud traced these curves in the mind to experiences of huge and frequently traumatizing emotional power, suppressed from conscious memory but still active at an unconscious level. They were revealed by situations or words which provoked distress, evasion or some other anomalous reaction in the patient. The goal of psychoanalysis, through an exchange of words between analyst and patient, was to trace the associations leading back from this reaction – be it a dream or a slip of the tongue – to the unconscious idea causing it. Freud's theory was that when someone brought a suppressed memory to light, or realized the hidden meaning of a dream, the disturbance in their mind would be resolved. If psychoanalysis did not smooth out the distortion in the psychic space-time of the mind, the person would at least understand why it was there.

In Freud's case, the crucial dream on the basis of which he formed his seminal hypothesis on the nature of dreams and the unconscious took place in July 1895. He dreamed about a young girl called Irma, a patient he was treating at the time. The dreams themselves are fairly complicated and in his writing Freud describes them extensively. To summarize, the story goes as follows: He meets Irma in a large ballroom. While other guests are arriving he talks to her and learns that she is suffering from a pain in her throat and her stomach. He worries that, as her doctor, he might have missed a physical (not mental) illness. After several other events take place he discovers that his friend Otto, also a doctor, had given Irma an infected injection, the true cause of her health issues.

Later Freud used a psychoanalytic method of free association to uncover the latent meaning of his dream. He remembered that, a day earlier, he had met Oscar Rie, a pediatrician who was also treating Irma. Rie told Freud that Irma was improving, but had not yet completely recovered. Freud took these words as a criticism of his work, suggesting moreover Oscar and not he was capable of treating Irma properly. Freud then immediately suppressed the feeling of wounded professional pride.

The method of treatment Freud in time developed into psychoanalysis always created a close bond between analyst and patient. In this case, at a stage when Freud was still unaware of the strength of that bond, Oscar's comments stirred powerful and possessive emotions in Freud. These Freud buried in his unconscious, unwilling to acknowledge how sensitive he was about the patient and her case, expressing them instead in the dream in which "Oscar" Rie became Otto, the poisoner of Irma, whose neurotic symptoms became physical. Through this "dream-work" Freud laid the blame for his sense of failure and guilt onto Otto/Oscar. In a way, the dream was a form of revenge against Oskar's nettling remarks, as Freud realized when he used free associations to map the uncomfortable curvature they had left in his mind. The cause he discovered for this warp lay hidden in the unconscious: inability to treat Irma's disease as a doctor triggered feelings of anxiety, so he had transferred this guilt onto another party, in this case the colleague who had seemingly criticized him.

The analogy between Einstein's time-space and Freud's unconscious is of course a loose one; but it can provide a tangible way of assessing the scale of Freud's momentous discovery. Even after a century the public perception of his work has often been based on poor and sometimes misleading interpretations.

Freud, who as a doctor and neurologist had a strong background in the natural sciences, gave considerable thought to the possible analogies his work in psychology suggested with other scientific endeavors. In one of his more technical works, intended mainly for colleagues, he wrote: "A psychoanalyst understands he is working with explosive forces and has to be as careful and meticulous as a chemist."

Why dessert comes last

When it comes to a great number of things, it is very difficult to imagine them not being done the way we always do them. Custom is so powerful that it feels as if our most deeply rooted habits are in fact something natural and predetermined. This is particularly true when it comes to how we prepare and consume our food. Many would be surprised to learn that the order of courses in a meal as well as the established combinations of tastes in an average European meal today have their origins in the middle of the seventeenth century. It was at that time that wealthier Europeans made significant changes to their eating habits, habits that have been taken for granted to this day. The reason for changing the established diet was a new scientific theory about how digestion works and what healthy nutrition really means.

Nature as one big kitchen

Since ancient times, people have been aware of the fact that their health depends greatly on the food they consume. During the High Middle Ages and Renaissance there was a rule defining the health of an individual as a state in which his bodily fluids coexisted in just the right proportions. Food was an important factor in maintaining this physical balance as doctors of the time had few other methods at their disposal. When someone was severely ill, they would reestablish the equilibrium of fluids within the body by controlled blood-letting. Only the wealthier part of the population, however, could afford such treatments, or to follow principles of what was thought at the time to be healthy nutrition.

According to an ancient medical tradition based principally on the works of the *Hippocratic Corpus*, Aristotle's essays and Galen's treatises, digestion was viewed as food being cooked inside the human body. Just as, in nature, the sun heats the earth so that plants grow from it and bear fruits, the inner flames of the human body were believed to "cook" ingested food further until it was transformed into bodily fluids, while the remains were disposed of in the form of excrements which would then fertilize the earth, and allow the natural cycle to repeat.

The resulting image of the cosmos as a big kitchen in which natural flames induce the growth of living nature and the circulation of matter was reflected in the image of the human body as a miniature version of this cosmic kitchen in which the inner flames cook the eaten food. According to this model of nature and man, the most frequent advice doctors gave their patients was that they should boil their food as much as possible, ensuring that their digestive mechanisms were put under less stress, and that they should consume balanced food. Of course, this balance should be viewed according to the way nutrients were classified at the time.

The theory of four elements

At the time, different types of food were classified according to Aristotle's established theory of the four elements from which the world was believed to have been created. Individual elements had been attributed corresponding characteristics. Fire was hot and dry, water was moist and cold, air was hot and moist, and earth was dry and cold. Food was classified in accordance with these four characteristics; pepper, for example, was put into the category of fire while melons, mushrooms and fish were classified under water. Because it was believed that the human body should ideally be moderately warm and moderately moist, meals were planned accordingly. A healthy diet meant eating food which came as close as possible to the ideal ratio of individual elements in the body.

The ideal meal was believed to be a sort of warm and relatively squishy porridge, because it contained all the characteristics that were supposedly best for the body. It is interesting that raw vegetables were, by the principles of the time, considered unhealthy and appropriate merely as food for the poor. On the other hand, sugar was believed to be a very healthy nutrient and an ideal food additive which was also sold in pharmacies. Because it was relatively expensive, it was usually kept safe or even locked up in the kitchen.

New science, new menu

In the 17th century, however, the diet of wealthier Europeans underwent a significant change. The ideals of nutrition had changed and so had advices for a healthy diet. In her books and articles, Rachel Laudan, a science historian researching the history of relations between science, medicine and nutrition, has come up with several convincing arguments which indicate that the change which occurred around 1650 in the eating habits of the rich did so mainly because of new scientific discoveries. The main reason for these changes was a new scientific theory about how human digestion worked and how, in nature, one substance changed into another. If earlier scholars had for the most part viewed processes in nature and the human body as being similar to cooking, their successors began to regard fermentation as the key natural phenomenon. In the most general terms, fermentation is a chemical process in which carbohydrates, like sugars, are transformed into alcohol and acids. With the fermentation of yeast, for example, one can produce wine, beer and vinegar. With the fermentation of lactic acid bacteria, milk changes into yogurt and other dairy products. Fermentation also holds the secret to how bread rises in the oven, causing the release of the carbon dioxide which leavens the dough and gives a baked loaf its typically bubbly structure.

The chemistry of tastes

The scholars who , a few centuries ago, studied the branch of science today known as chemistry, used fermentation and distillation to isolate individual key constituents s of plants in the search for new healing ingredients. Because fermentation was present throughout nature as well, the theory was quickly established that human digestion did not involve cooking the food the stomach, but rather in fermenting it.

The system of food classification was quickly adapted to the new theory of digestion. Aristotle's four elements were replaced by a system of classification based on three new ideals of pure substances which were then first introduced in chemistry. The three ideal substances were salt, oil and quicksilver. However, these are not to be understood in the modern sense of the words, but in a more (al)chemical sense, pertaining to the roles these substances assumed in the processes of fermentation and distillation. The chemists of the time discovered that substances in distillation substances usually separate into three parts: volatile liquids, "oily" substances and solid residues.

Sugar becomes poison

"Oil" was the term used to designate substances that did not evaporate during the boiling process, in contrast to alcohol which was classified as one of the "quicksilver" substances. Among the food items that could be found in the kitchen, butter, lard and olive oil were the closest to the "element of oil". Apart from regular salt, flour and similar solid substances were part of the "salt" group. The typical representatives of the "quicksilver" group were vinegar, wine and other alcoholic beverages, as well as certain flavors of meat and fish. With the new division of food, oils especially gained in importance, becoming key ingredients of a variety of sauces. The new food classification system caused a noticeable change in eating habits.

As for sugar, doctors began to realize that it was far from the ideal nutrient it was once believed to be. They discovered, for example, that it damaged teeth, and was also to be found in the urine of patients suffering from what was later defined as diabetes. According to the new division of nutrients, sugar was no longer believed to be an ideal type of food: in fact some doctors came close to describing it as poison. That is why for the last few centuries our main courses have no longer contained excessive amounts of sugar, and sweets have only been consumed in small quantities at the end of a meal.

A NOTE ON THE AUTHOR

Sašo Dolenc (*saso-dolenc.com*) is a physicist and philosopher by training. He is cofounder and editor in chief of the on-line popular science journal Kvarkadabra (*www.kvarkadabra.net*), whose mission is to explain science in a simple, attractive and fun form that is open to all.

The aim of the articles collected in the series of *Short Stories from Science, History and Philosophy* is to present some key scientific events, ideas and personalities in the form of short tales that are easy and enjoyable to read. Scientific and philosophical concepts are explained in a way that anyone may understand. Each story may be read separately, but at the same time they all band together to form a wide-ranging introduction to the history of science and areas of contemporary scientific research, as well as some of the recurring problems science has encountered in history and the philosophical dilemmas it raises today.

In Praise of the Series

"Each chapter in this book explains a complex problem through moving, amusing and marvellous stories. Sašo Dolenc's recurring theme is the elusive and often eccentric nature of inspiration; but in exploring it he covers an immense variety of subjects, from meteorology to microbiology, computer technology to market theory. His readers will gain a succinct and satisfying lesson on each topic, and a sense overall of the labour, genius and luck that science demands."

– **John Stubbs**, author of *John Donne: The Reformed Soul* and *Reprobates: The Cavaliers of the English Civil War*

"Great fun. Like Malcolm Gladwell, Dolenc writes about complicated science in a clear, accessible way that entertains and educates. The smarter and better the writer, the clearer and simpler he will make concepts that are difficult to grasp. Reading this book is a pleasure you can learn from."

– **Noah Charney**, best-selling author of *The Art Thief* and *Stealing the Mystic Lamb*

"If I were the only survivor on a remote island and all I had with me were this book, a Swiss army knife and a bottle, I would throw the bottle into the sea with the note: 'Don't worry, I have everything I need.'"

– Ciril Horjak, alias **Dr Horowitz**, a comic artist

"Does science think? Heidegger once answered this question with a decisive No. The writings on modern science skillfully penned by Sašo Dolenc, these small stories about big stories, quickly convince us that the contrary is true. Not only does science think in hundreds of unexpected ways, its intellectual challenges and insights are an inexhaustible source of inspiration and entertainment. The clarity of thought and the lucidity of its style make this book accessible to anyone ... in the finest tradition of popularizing science, its achievements, dilemmas and predicaments."

— **Mladen Dolar**, philosopher and
author of *A Voice and Nothing More*

"Sašo Dolenc is undoubtedly one of our most successful authors in the field of popular science, possessing the ability to explain complex scientific achievements to a broader audience in a clear and captivating way while remaining precise and scientific. His collection of articles is of particular importance because it encompasses all areas of modern science in an unassuming, almost light-hearted manner."

— **Boštjan Žekš**, physicist and former president
of the *Slovenian Academy of Sciences and Arts*

"The writing is understandable, but never simplistic. Instructive, but never patronizing. Straightforward, but never trivial. In-depth, but never too intense."

— **Ali Žerdin**, features editor at *Delo*,
the main Slovenian newspaper

Made in the USA
Monee, IL
05 November 2026

67811529R00174